Progress in IS

More information about this series at http://www.springer.com/series/10440

David Geiger

Personalized Task Recommendation in Crowdsourcing Systems

Springer

David Geiger
University of Mannheim
Mannheim
Germany

This book is based on a doctoral thesis submitted to the University of Mannheim.

ISSN 2196-8705 ISSN 2196-8713 (electronic)
Progress in IS
ISBN 978-3-319-22290-5 ISBN 978-3-319-22291-2 (eBook)
DOI 10.1007/978-3-319-22291-2

Library of Congress Control Number: 2015950152

Springer Cham Heidelberg New York Dordrecht London

Printed on acid-free paper

Springer International Publishing AG Switzerland is part of Springer Science+Business Media (www.springer.com)

Abbreviations

API	Application programming interface
CI	Confidence interval
CSS	Cascading Style Sheets
DDD	Domain-driven design
DOM	Document Object Model
HIT	Human Intelligence Task
HTML	Hypertext Markup Language
HTTP	Hypertext Transfer Protocol
IETF	Internet Engineering Task Force
IP	Internet Protocol
IS	Information systems
JAX-RS	Java API for RESTful Web Services
JS	JavaScript
JSON	JavaScript Object Notation
REST	Representational State Transfer
RFC	Request for Comments
UI	User interface
URI	Uniform Resource Identifier
URL	Uniform Resource Locator
W3C	World Wide Web Consortium

Contents

Figures

Tables

Chapter 1
Introduction

Crowdsourcing is an umbrella term for approaches that harness the diverse potential of large groups of people via an open call for contribution over the Web. Using information technology as a facilitator, crowdsourcing organizations implement socio-technical systems to channel the contributions of human workforce, knowledge, skills, or perspectives within and beyond organizational boundaries into the generation of digital information products and services. Crowdsourcing systems have recently gained in popularity for a multitude of organizational functions, such as problem solving, knowledge aggregation, content generation, and large-scale data processing.

1.1 Research Context

All crowdsourcing systems are built on the same fundamental principle of contributor self-selection: Organizations, often called requesters, publish their tasks to a typically unrestricted public on the Web and interested individuals decide whether they are willing and able to contribute. As opposed to traditional employee models, this approach enables crowdsourcing organizations to implement tasks that benefit from the exposure to a wide variety of potential contributors with diverse backgrounds. Building a successful crowdsourcing system hence entails the challenge of reaching and enlisting a sufficient number of individuals with suitable interests and capabilities.

Research into contributor motivation across crowdsourcing systems has shown that task preferences are subject to a complex multidimensional construct of motivational factors, which are weighted differently across individuals (Antikainen et al. 2010; Brabham 2008, 2010; Kaufmann et al. 2011; Leimeister et al. 2009; Schroer and Hertel 2009; Schulze et al. 2012; Sun et al. 2012; Zheng et al. 2011). While the findings confirm the role of extrinsic factors such as payment, they highlight in particular the significance of intrinsic motivation, i.e., "doing

D. Geiger, *Personalized Task Recommendation in Crowdsourcing Systems*,
Progress in IS, DOI 10.1007/978-3-319-22291-2_1

something because it is inherently interesting or enjoyable" rather than "doing something because it leads to a separable outcome" (Ryan and Deci 2000, p. 55). This means that an individual's task preferences are not merely based on explicit and often task-agnostic incentives set by the task requester. Instead, they depend to a considerable extent on the match of intrinsic personal interests with the very nature, i.e., the type or topic, of a task.

The capabilities required to make a potentially valuable contribution to a task often take the form of specific knowledge or skillsets. The more specific a task and its requirements are, the more limited is the group of potential contributors. Required capabilities may be specified explicitly by the task requester or they may result implicitly from the nature of the task. Some capabilities can be formalized as "qualifications" (Lopez et al. 2010; Paolacci et al. 2010) that potential contributors need to obtain before they are allowed to contribute to certain tasks, for instance, by passing a corresponding test. In many cases, however, requesters need to rely on the contributors' perceived compliance with the explicit or implicit requirements.

While the self-selection principle enables interested individuals to contribute to tasks that match both their personal interests and capabilities, current systems do not provide adequate support in finding such tasks among the vast opportunities of engaging in crowdsourcing activities. The rising popularity of crowdsourcing approaches has led to a strongly increasing number of projects, which require contributors to perform a diverse set of tasks and roles (Zhang et al. 2011). In October 2014, crowdsourcing.org, a leading industry directory, maintains a list of over 2800 crowdsourcing examples in a plethora of application domains. Among these examples are extensive collective efforts such as Wikipedia and marketplaces that act as intermediaries for thousands of crowdsourcing organizations. Without the assistance of efficient mechanisms, individuals can hardly navigate the massive body of tasks accumulated by these systems and easily suffer from information overload. Indeed, several studies have shown that potential contributors struggle to find those tasks that they are strongly willing and capable of participating in.

A prominent example to observe the difficulty of finding suitable tasks is the well-established Mechanical Turk marketplace. Mechanical Turk lists a variety of 'human intelligence tasks' (HITs) that draw on typical human abilities and cognitive skills, such as natural language processing or providing personal opinions (Ambati et al. 2011; Carignani and Lorenzo 2010). For the period between July and December 2014, the Mturk-Tracker (Ipeirotis 2010) on average shows over 2000 open tasks ('HIT groups'), comprising about 350,000 individual assignments. The list of HIT groups displays ten tasks per page and can be sorted by generic criteria such as 'HIT creation date' or 'HITs available,' i.e., the number of remaining assignments, also called task instances. Interested individuals need to register an account with Mechanical Turk, browse and select from the list of tasks, and eventually submit their contributions (Corney et al. 2009). Studies of the contributor population, typically called 'workers,' have shown a wide variance in demographic factors such as gender, age, income, and education (Ross et al. 2010).

In a survey on the impact of task properties, Mechanical Turk workers reported "HIT sounds interesting/enjoyable," i.e., an intrinsic interest, among the highest-rated criteria for choosing a task (Schulze et al. 2011). An analysis of task disappearance rates for different sorting criteria, however, find that workers mostly choose from "the first page of the most recently posted tasks and the first two pages of the tasks with the most available instances" (Chilton et al. 2010). As these sorting criteria do not relate to the nature of a task, workers seem to make a trade-off with respect to search costs, which results in a suboptimal match with personal interests and capabilities. This is confirmed by the results of a corresponding survey, in which about 25 % of the workers state that they occasionally go to great lengths and browse dozens of pages in order to find suitable tasks. Several workers specifically complain about the difficulty of finding "good" tasks, which clearly shows the need for better mechanisms (Chilton et al. 2010).

Similar results have been presented by studies of other crowdsourcing systems. Cosley et al. (2007) find that member-maintained communities, such as Slashdot or Wikipedia, do not provide their contributors with sufficient help to find work and fail to consider individual preferences. Dror et al. (2011) determine that "users of Yahoo! Answers find it hard to search for questions they would like to answer" and "mainly skim through long and dynamic lists of open questions." Stankovic et al. (2011b) conclude from their practice at Hypios, a problem solving intermediary, that "simply posting a problem online is not enough to attract diverse and innovative solutions," but that platforms need to proactively identify potential solvers. Jeppesen and Lakhani (2010) show that successful solutions are frequently provided by contributors that come from lateral domains and can offer different perspectives on a problem; a phenomenon also called "serendipity" (Stankovic et al. 2011a). Many individuals, however, will only look at tasks in their primary field of expertise, not realizing on their own that they possess the right capabilities to solve tasks from other fields (Jesic et al. 2011).

1.2 Research Objective

In order to assist potential contributors with the discovery of suitable tasks according to their individual interests and capabilities, this book explores the design of mechanisms for personalized task recommendation. Recommender systems are a category of tools and techniques that provide suggestions of potentially useful items based on individual preferences (Ricci et al. 2011). They have achieved enormous success in many areas with similar challenges of information overload, such as product recommendation in e-commerce or content personalization in information and entertainment portals. Note that a few recommender systems also recommend items based on general characteristics, e.g., popularity, but the concept usually refers to techniques that make recommendations based on a model of individual user preferences. In line with the literature, this book therefore uses the terms *recommender system* and *personalized recommender system* interchangeably.

Little research has addressed the use of recommender systems in a crowdsourcing context. Although a few authors have mentioned the challenge of improving the alignment of contributors and tasks in specific contexts (Alonso 2011; Ipeirotis 2010; Vukovic 2009), current crowdsourcing systems do not adequately address this need. In most cases, the corresponding design features are limited to standard search functionality as well as task sorting and filter mechanisms that are based on generic criteria such as thematic category, payment, or date. Unlike recommender systems, such criteria cannot appropriately cover the inherent characteristics of the available tasks; nor can they provide a sufficient level of personalization to the individual contributor.

By facilitating the match of personal interests and capabilities with the right tasks, a recommendation approach will create benefits for both contributors and requesters. Contributors that do not need to invest high search costs or settle for suboptimal tasks are likely to maintain a higher motivation. Yuen et al. (2011) argue that more contributors could be attracted in the long run if they found suitable tasks more easily. Cosley et al. (2007) show that theory in economics and social psychology suggests that "reducing the cost and increasing the personal value of contribution would motivate members to participate more." Their argument is partly based on the collective effort model, which states that personal value depends, among others, on the individual "meaningfulness and intrinsic value of the task" (Karau and Williams 1993). Furthermore, the model predicts that an individual's motivation to contribute to collective tasks is higher if the contributions are perceived as unique. Intuitively, this applies to those crowdsourcing systems that emphasize collaboration among contributors, as opposed to competition, and the effects have been confirmed in such contexts (Ling et al. 2005; Ludford et al. 2004). A recommendation that highlights the value of individual capabilities to a particular task might induce such a perception of uniqueness (see Cosley et al. 2006).

From a requester perspective, task recommendation increases the probability of attracting contributions from individuals that are highly motivated and possess the right capabilities. For tasks with very specific requirements, such as in crowd solving systems, this can be critical to success. As a result of the information overload, requesters currently miss out on many good contributions because their tasks are often not discovered by the most capable individuals. Moreover, less suited contributors, potentially motivated by monetary incentives, may submit contributions that result in sub-standard or noisy output instead (Ambati et al. 2011; Mason and Watts 2009). In systems with a fixed number of task assignments, such as Mechanical Turk, this may even prevent contributions from truly skilled individuals who have not been able to find the tasks in time (Ambati et al. 2011). Some systems seek to counter these adverse effects by employing qualification-based restrictions of potential contributors (Geiger et al. 2011b) or crowd-based mechanisms to assure the validity of homogeneous contributions, such as majority votes or peer reviews (Kern et al. 2010, 2012). These measures typically further extend the cost and the duration of a task. In summary, personalized task recommendation mechanisms have a considerable potential to improve the overall efficiency and the outcome of crowdsourcing approaches (Ho and Vaughan 2012; Yuen et al. 2011).

1.3 Research Approach

This book presents a series of consecutive studies that were conducted to pursue the described objective. While the individual studies employ a variety of research methodologies, the overall knowledge building process is fundamentally based upon the design science research paradigm (Hevner and Chatterjee 2010; Hevner et al. 2004; Peffers et al. 2007). In a first step, Chap. 2 lays out a conceptual framework for research into the crowdsourcing phenomenon from a socio-technical systems perspective. The framework distinguishes the various functions that crowdsourcing systems assume within an organizational environment and describes the nature of the involved components. Chapter 3 then analyzes the current state of research on personalized task recommendation in crowdsourcing systems. To this intent, a systematic literature review is conducted, focusing on the context, the techniques, and the evaluation of existing studies. Building on the insights derived from this review, Chap. 4 describes the design of a new, third-party task recommendation service that integrates with existing crowdsourcing platforms. The service enhances real-world systems with personalized task recommendation capabilities and thus also enables corresponding field research in productive environments. In order to evaluate the design, Chap. 5 describes a series of three studies that were conducted along the process of deploying the prototype on the Mechanical Turk platform. In particular, the studies provide empirical evidence on the acceptance and utility of personalized task recommendation from a contributor perspective. Finally, Chap. 6 concludes this book with a summary of the contributions and an outlook on potential future research.

Chapter 2
Crowdsourcing Systems

While isolated examples for crowdsourcing approaches can be found throughout the centuries (Surowiecki 2005), the development of the Internet and Web 2.0 technologies has drastically leveraged the reach and the efficiency of connecting with large groups of people. Ever since, researchers and practitioners from various fields have begun to explore how to tap into the potential of crowds in a multitude of contexts.

It is only recently that such approaches have been considered as related phenomena. The term "crowdsourcing" (Howe 2006) has since attracted a lot of attention, even if its exact understanding still varies (Estellés-Arolas and González-Ladrón-de-Guevara 2012). Relevant research is rooted in fields as diverse as computer science (Doan et al. 2011) and management (Schenk and Guittard 2011) and is applied in many other areas that have discovered crowdsourcing as a potentially useful approach. Consequently, research questions mostly center on specific use cases and individual aspects of crowdsourcing. This fragmentation of the research landscape makes it difficult to establish a comprehensive knowledge base and to provide structured guidance for the design of crowdsourcing solutions.

Inspired by systems theory, this chapter introduces a more integrated, socio-technical perspective to study the information systems that realize Web-based crowdsourcing efforts. The resulting framework provides a conceptual structure to channel and relate crowdsourcing research. In particular, it can be used to gain a deeper and comparative understanding of different types of crowdsourcing systems and thus guide the systematic analysis and design of their components, e.g., with respect to task recommendation.

The following sections summarize the product of a series of previous studies (Geiger et al. 2011a, b, 2012). Section 2.1 first establishes a socio-technical definition of crowdsourcing systems and their individual components. Section 2.2 then proceeds to analyze the organizational functions of crowdsourcing systems, identifying four system archetypes with distinct characteristics. Section 2.3 briefly discusses the usefulness of the resulting typology.

D. Geiger, *Personalized Task Recommendation in Crowdsourcing Systems*,
Progress in IS, DOI 10.1007/978-3-319-22291-2_2

2.1 A Socio-Technical Perspective

A system is a set of interrelated elements or components that work together to achieve an overall objective. Systems have a clearly defined boundary and exist as components or subsystems of other systems, also called the environment of a system. Most systems are open, i.e., they interact with their environment via interfaces. Systems are ubiquitous and can, for instance, be of biological, technical, or social nature (Ackoff 1971; Bertalanffy 1972; Churchman 1968).

Information systems (IS) are subsystems of an organizational system that provide an organization with information services needed for operations and management (Davis 2000; Falkenberg et al. 1998; Heinrich et al. 2011). Understandings of the term *information system* differ widely in the extent to which they emphasize social vs. technical concerns (Alter 2008; Carvalho 2000; Falkenberg et al. 1998). While some academics hold a primarily technical view (e.g., Ein-Dor and Segev 1993), the majority of the IS community view information systems as socio-technical systems that integrate human and machine components (Davis 2000; Heinrich et al. 2011; Kroenke 2011; Land 1985; Valacich et al. 2011; WKWI 1994). Well-founded socio-technical approaches study information systems within their organizational context and hence ensure that the individual elements, e.g., the IT component, and their design are aligned with this context and with each other (Alter 2008; Carvalho 2000; Lyytinen and Newman 2008).

As one particular socio-technical approach, the work system approach defines an information system as "a system in which human participants and/or machines perform work (processes and activities) using information, technology, and other resources to produce informational products and/or services for internal or external customers" (Alter 2008, p. 451). Informational products are understood in a broad sense and include, for instance, the production of digital goods. The work that is performed in such a system, i.e., its processes and activities, is devoted to a generic function of processing information, which involves "capturing, transmitting, storing, retrieving, manipulating, and displaying information" (Alter 2008, p. 451). As a means to describe and analyze an information system, the work system framework identifies, among others, the four basic components that are involved in performing the work: processes and activities, (human) participants, information, and technologies. The work system view of the function and elements of information systems can easily be mapped to the definitions in most of the IS textbook literature (Davis 2000; Ferstl and Sinz 2008; Gray 2005; Heinrich et al. 2011; Huber et al. 2006; Kroenke 2011; Rainer and Cegielski 2010; Valacich et al. 2011).

Building on this definition, *crowdsourcing systems* can be perceived as socio-technical systems that provide informational products and services for internal or external customers by harnessing the diverse potential of large groups of people, usually via the Web. All of these systems employ variations of a generic crowdsourcing process that relies primarily on contributions from human participants to transform existing or produce new information. Information technology is used to enable this process and, where possible, support the activities performed in the system. Figure 2.1 summarizes the nature of the main components in a crowdsourcing system.

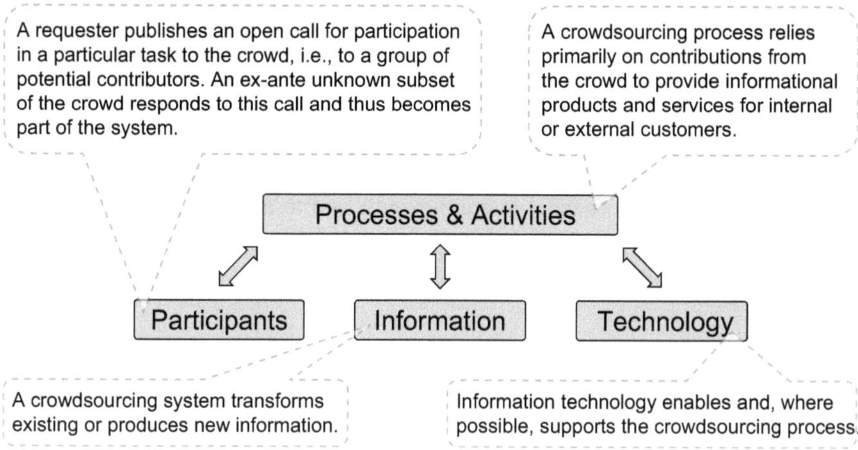

Fig. 2.1 Components of a crowdsourcing system

A crowdsourcing process starts with a crowdsourcing organization, often called requester, that publishes an open call for participation in a particular task to the *crowd*, i.e., to a typically unrestricted group of potential contributors. In response to this call, an ex-ante unknown subset of individuals in the crowd decide if they are willing and capable of contributing to the respective task. This essential self-selection part of the process is also called "self-identification of contributors" (Howe 2009). Individuals that choose to participate submit their contributions, which are then aggregated and selected in different ways (Schenk and Guittard 2011), depending on the function of the crowdsourcing system.

The open nature of their processes enables crowdsourcing systems to reach a large number of potential contributors and, therefore, to scale very well. At the same time, interested individuals may come from entirely different backgrounds, which creates the potential for a high level of diversity in the respective crowd. It is this unique scalability and diversity that, if implemented correctly, ensure the large success of crowdsourcing systems for a variety of organizational functions.

2.2 Organizational Functions

According to general systems theory, a system cannot be understood by analyzing its constituent parts alone, but only by determining its function in the supersystem, i.e., the surrounding system of which it is a part (Ackoff 1993; cited in Silver et al. 1995). Like any information system, a crowdsourcing system thus needs to be considered in terms of its function within the organizational environment before one can derive details on its "features and component parts" and how they enable this function (Silver et al. 1995).

Consequently, in order to facilitate a better understanding of the respective socio-technical designs, this section lays out a distinction of fundamental crowdsourcing system archetypes along their organizational function. The development of typologies provides structures to organize the body of knowledge and to study relationships among otherwise disorderly concepts (Glass and Vessey 1995; Nickerson et al. 2013). By enabling a more differentiated perspective, a typology of systems can be used to channel research results and to attribute divergences, e.g., with respect to the design of task recommendation approaches, to systematic differences (Sabherwal and King 1995).

2.2.1 Typology Development

Nickerson et al. (2013) describe a structured approach to the development of classification schemes, which is based on a comprehensive literature survey of 73 taxonomies in IS and on methodological guidance from related fields. Note that, according to Nickerson et al., the terms *taxonomy* and *typology* are often used interchangeably and can refer to both a set of characteristics and a set of classified objects. One of the main traits of the method is the definition of a meta-characteristic as a first step to the development of any taxonomy:

> The meta-characteristic is the most comprehensive characteristic that will serve as the basis for the choice of characteristics in the taxonomy. Each characteristic should be a logical consequence of the meta-characteristic. The choice of the meta-characteristic should be based on the purpose of the taxonomy. (Nickerson et al. 2013, p. 8)

The purpose of this typology, as stated above, is to distinguish archetypes of crowdsourcing systems along their organizational function. Following the work system definition, the general function of any information system can be perceived as the processing of information in order to provide informational products and services (Alter 2008). As opposed to traditional information systems, the product or service that is provided by a crowdsourcing system is essentially determined by the input of its various human contributors. The most comprehensive characteristic to differentiate between the organizational functions of these systems therefore considers how a system derives value from crowd contributions to deliver the aspired result.

As prescribed by the method, an iterative combination of empirical-to-conceptual and conceptual-to-empirical approaches was employed to develop the typology. On the empirical side, this process involved analyzing a sample of over 50 crowdsourcing systems. On the conceptual side, inspiration was drawn from concepts used in systems theory and from existing attempts at categorizing the crowdsourcing landscape in academia and industry. The iterative approach involved a broad consideration of candidate dimensions, which were tested for their individual relevance to the meta-characteristic and their collective distinctive potential. Nickerson et al. provide a number of objective and subjective criteria that serve as both ending conditions for the development process and evaluation criteria for the resulting artifact. The next section presents this artifact, followed by a brief discussion of its compliance with the evaluation criteria.

2.2.2 System Archetypes

The developed typology covers the specified meta-characteristic—how a crowdsourcing system derives value from crowd contributions to achieve its organizational function—by differentiating two fundamental dimensions: (i) whether a system seeks homogeneous or heterogeneous contributions from the crowd and (ii) whether it seeks an emergent or a non-emergent value from these contributions.

i. A system that seeks *homogeneous* contributions values all valid contributions equally. Homogeneous contributions that comply with the predefined specifications are seen as qualitatively identical; the system is geared to mere quantitative processing. In contrast, a system that seeks *heterogeneous* contributions values these contributions differently according to their individual qualities. Relevant qualities are determined by the specific task and vary from objective measures, such as test results, to subjective perceptions, such as clarity or esthetics. Heterogeneous contributions are seen as alternatives or complements and are processed accordingly. This dimension is inspired by the notion of heterogeneous components (or components perceived as such), which is studied in various systems (Heinrich et al. 2011, p. 16). A particular focus on heterogeneity can be found, for instance, in the context of distributed computer systems (Maheswaran et al. 1999) or agent models in economic systems (Hommes 2006).
ii. A system that seeks a *non-emergent* value from its contributions derives this value directly from all or some of the individual contributions in isolation. In such systems, an individual contribution delivers a fixed value, which is independent of other contributions. A system that seeks an *emergent* value from its contributions, however, can only derive this value from the entirety of contributions and the relationships between them. An individual contribution therefore only delivers value in combination with others. Emergence is a philosophical concept that is, among others, central to systems theory to denote properties of a system that are not possessed by its isolated components but rather depend on the relationships among them in a composition (Bunge 2003, p. 12 ff.; Checkland 1988, p. 243; Heinrich et al. 2011, p. 15; Weber 1997, p. 37).

The combination of these two dimensions yields four fundamental types of crowdsourcing systems as depicted in Fig. 2.2. Each of the archetypes was given a label that reflects its distinct organizational function and, accordingly, the type of product or service delivered by the system.

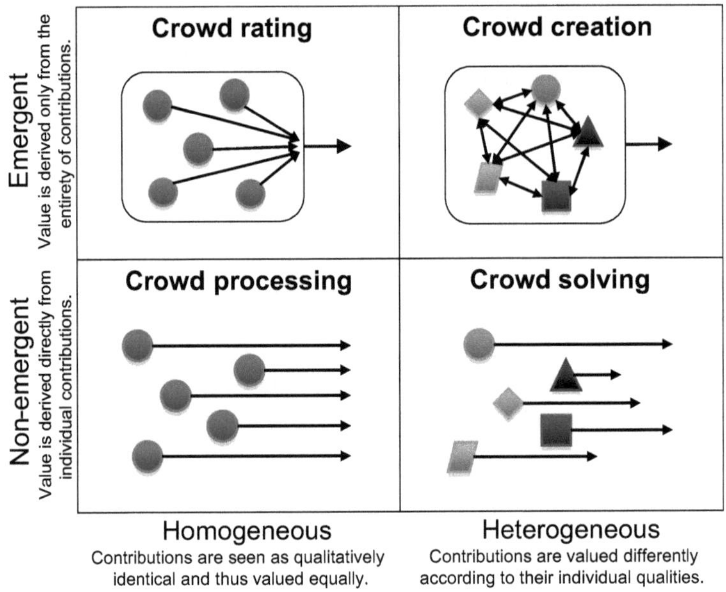

Fig. 2.2 The four archetypes of crowdsourcing systems

Crowd processing systems seek non-emergent value that derives directly from large quantities of homogeneous contributions. Valid contributions represent qualitatively identical and therefore equally valued chunks of work. These systems utilize the additional bandwidth and the scalability provided by a crowdsourcing approach for quick and efficient batch processing, thus minimizing the use of traditional organizational resources (Doan et al. 2011). The diversity of interests within the crowd enables these systems to recruit sufficient contributors for a variety of tasks, most of which harness basic abilities of the human brain. Prominent examples for crowd processing systems include Mechanical Turk,[1] Galaxy Zoo[2] (Lintott et al. 2008), and Recaptcha[3] (von Ahn et al. 2008).

Crowd rating systems also rely on large amounts of homogeneous contributions but seek a collective value that emerges only from the entirety of contributions. As homogeneous contributions are considered qualitatively identical, the aspired value emerges from the mere quantitative properties of the collective input. Contributions hence represent 'votes' on a given topic. By aggregating a sufficient number of these votes, crowd rating systems deduce a collective response, such as a collective assessment on TripAdvisor[4] or a collective prediction on the Hollywood Stock

[1] http://www.mturk.com/

[2] http://www.galaxyzoo.org/

[3] http://www.google.com/recaptcha/

[4] http://www.tripadvisor.com/

Exchange.[5] Typically, the larger and more diverse a crowd that rating systems can assemble, the more accurate the results become. In fact, Surowiecki (2005) states diversity as one of the prerequisites for the "wisdom of crowds" to emerge.

Crowd solving systems seek non-emergent value that derives directly from the isolated values of their heterogeneous contributions. Contributions are qualitatively different and thus represent alternative or complementary solutions to a given problem. These systems can be built around hard problems with well-defined evaluation criteria, such as FoldIt[6] (Cooper et al. 2010), kaggle[7] (Carpenter 2011), or the Netflix Prize (Bennett and Lanning 2007). They can also be used to approach soft problems that do not have an optimal solution, such as ideation contests on InnoCentive[8] (Allio 2004) or ideaBounty[9] and 'make-to-order' digital product contests on 99designs[10] or Naming Force.[11] Crowd solving systems benefit from large and diverse crowds as every individual potentially provides new insights, ideas, or abilities and therefore increases the chance of finding a (better) solution. In some contexts, this phenomenon is referred to as the "wisdom in the crowd" (Dondio and Longo 2011, p. 113).

Crowd creation systems seek a collective value that emerges from the accumulation of a variety of heterogeneous contributions. In contrast to crowd rating systems, the aspired value emerges not only from quantitative but primarily from qualitative properties of the collective input. Contributions have a complementary share in the collective outcome depending on their individual qualities and their relationship with others. Large numbers of diverse contributors enable crowd creation systems to harness multiple perspectives, distributed knowledge, or different skills, and to aggregate the corresponding contributions into highly comprehensive artifacts. Typical examples are user-generated content platforms (e.g., YouTube[12]), the 'make-to-stock' production of digital content (e.g., stock photography), or knowledge aggregation (e.g., Wikipedia[13]).

As stated above, the described systems are archetypes with distinct organizational functions. In practice, many crowdsourcing efforts are built on hybrid systems that combine some of these functions, often relying on quantitative and qualitative components. Many systems that are based on heterogeneous contributions, for instance, rely on a crowd rating function in form of a collective vote as an

[5] http://www.hsx.com/

[6] http://fold.it/

[7] http://www.kaggle.com/

[8] http://www.innocentive.com/

[9] http://www.ideabounty.com/

[10] http://99designs.com/

[11] http://www.namingforce.com/

[12] http://www.youtube.com/

[13] http://www.wikipedia.org/

indicator for the quality of individual contributions. Some examples include Dell's IdeaStorm,[14] iStockphoto,[15] or Google's App Store.[16] Additionally, some of these systems make use of crowd processing functions such as tagging in order to organize the set of input elements.

2.3 Discussion

According to the design science paradigm, a "search for the best, or optimal, design is often intractable for realistic information systems problems" and should instead aim to discover effective solutions (Hevner et al. 2004, p. 88). On this basis, Nickerson et al. (2013) argue that taxonomies can only be evaluated with respect to their usefulness. They propose a set of qualitative attributes that form the necessary conditions for a useful taxonomy: it needs to be concise, robust, comprehensive, extendible, and explanatory.

For the scope of this work, the developed typology satisfies these criteria. The number of characteristics is *concise* enough to be easily applied, yet they provide a *robust* differentiation of distinct system archetypes. The typology is *comprehensive* because its disjoint characteristics partition the set of crowdsourcing systems, which allows the classification of every system in the used sample. While future work could *extend* the typology by identifying further subtypes, the current version has sufficient *explanatory* power with respect to the essential mechanisms and the organizational functions of the classified system instances.

In addition to these generic, necessary conditions, however, Nickerson et al. note that the sufficient conditions for usefulness depend on the expected use of a specific taxonomy. As classification schemes are not an end in themselves, their usefulness can only be evaluated by observing their use over time, with regards to their respective purpose. Following that idea, the next chapter proceeds to apply the typology to the comparative study of personalized task recommendation approaches in crowdsourcing systems.

[14] http://www.ideastorm.com/

[15] http://www.istockphoto.com/

[16] http://play.google.com/store

Chapter 3
Current State of Personalized Task Recommendation

In order to build a foundation for the design of mechanisms that assist individuals in finding suitable tasks, this chapter describes a review of the related academic literature. By channeling any existing efforts into a systematic perspective, this review generates a comprehensive picture of the current state of personalized task recommendation and provides insights that exceed the boundaries of individual crowdsourcing systems.

The chapter is laid out as follows. Section 3.1 describes the details of the chosen methodological approach. Section 3.2 proceeds with a presentation of the corresponding findings, focusing on the context, the techniques, and the evaluation of current research. Based on these findings and well-established recommender systems literature, Sect. 3.3 discusses a number of key issues for future research on the design of task recommendation approaches. A previous version of this study was published in the *Journal of Decision Support Systems* (Geiger and Schader 2014).

3.1 Methodology

The systematic literature review described in this chapter follows the methodological guidelines described by Webster and Watson (2002) and Levy and Ellis (2006). This section briefly summarizes how these guidelines were implemented to accumulate a body of literature. It also elaborates on the used search parameters.

The literature search was not restricted to specific outlets or disciplines for several reasons. Firstly, crowdsourcing has only recently been identified as an overarching phenomenon (Howe 2009) and has as such aroused the interest of academic researchers. As a consequence, the vast majority of crowdsourcing articles has not yet found its way into high quality journals and, instead, is published at (peer-reviewed) conferences. Secondly, crowdsourcing is a socio-technical approach with many potential applications. Corresponding research thus comes

D. Geiger, *Personalized Task Recommendation in Crowdsourcing Systems*,
Progress in IS, DOI 10.1007/978-3-319-22291-2_3

from an increasing variety of fields studying how to tap into the potential of crowds. Some of these fields, e.g., computer science, also attribute a higher value to conferences as primary publication outlets (Freyne et al. 2010). Finally, initial scans of the literature indicated that the body of knowledge with respect to the emerging topic of personalized task recommendation is very limited. Consequently, this review pursues an inclusive approach that aims to identify and derive insights from the few existing studies in order to provide a foundation for future research.

To this intent, an extensive keyword search was conducted on Google Scholar. The search engine was chosen due to its comprehensive coverage across the boundaries of individual databases. On the downside, this approach tends to generate a large number of irrelevant results, particularly when conducting isolated searches for terms that are used in a variety of contexts, such as 'recommender system.' In line with the object of this study, the literature search therefore employed a combination of keywords from two groups: (i) terms that refer to crowdsourcing approaches, such as 'crowdsourcing,' 'crowd,' 'crowdsourced,' or 'community,' and (ii) terms that cover the recommendation aspect, such as 'task recommendation,' 'task routing,' 'task matching,' 'recommender system,' or 'collaborative filtering.' The keyword selection was refined iteratively based on the results of the search queries.

Search results on Google Scholar were sorted by a combination of textual match and academic impact as measured, e.g., by citations. Potentially relevant articles were identified by scanning the titles and abstracts of the search results for indicators of task recommendation in crowdsourcing. Several hundred results for the various keyword combinations were scanned until no further relevant articles could be identified. The identified articles were then read in detail to determine whether they actually described, developed, or evaluated an approach for matching tasks with contributors in crowdsourcing systems as defined in Sect. 2.1.

A subsequent step involved backward and forward searches (Webster and Watson 2002) on the accumulated body of literature to identify further articles and to ensure no major contributions were missed. As Levy and Ellis (2006) note, keywords of IS literature tend to change regularly, especially such fairly recently introduced 'buzzwords' as crowdsourcing. Hence, this step also enabled the identification of further, similar research streams that are not termed *crowdsourcing* but are based on the same mechanics.

The search process ended when no more relevant articles could be identified by extending the search with additional keywords or by following cross references. This is in line with the ending conditions described in Webster and Watson (2002) as well as in Levy and Ellis (2006). Despite all efforts, a literature review can never be completely exhaustive, especially if the topic of interest has potentially been studied in many different contexts. Nevertheless, the described approach will have identified the bulk of the relevant literature and result in an accurate depiction of the current state of the art.

The accumulated body of literature comprises a total of 27 articles. As predicted, few of the articles have been published in journals (Jamjoom et al. 2009; Liu and Hao 2012; Yuan et al. 2009) and the vast majority has been published in conference proceedings. Non-published articles such as working papers and term papers as well as duplicate publications were excluded. One paper that could not be accessed was also excluded (Ni et al. 2012). Another study did not provide sufficient details about the recommendation approach (Kulkarni et al. 2012). Table A.1 gives an overview of the identified literature and the respective outlets.

3.2 Findings

The following sections summarize the findings regarding the emerging state of personalized task recommendation in crowdsourcing systems. The analysis focuses on the context of current research, the techniques used to generate personalized recommendations, and the evaluation of existing approaches. Note that some of the identified articles do not provide insights on all three aspects and therefore do not appear in all subsections.

3.2.1 Recommender Context

As a first level of analysis, the context of the existing studies was considered. The findings show that research on personalized task recommendation is conducted in various streams that remain largely isolated from each other. As described in Sect. 2.2, current approaches were classified with respect to the organizational function that the underlying crowdsourcing system addresses. In contrast to a large part of the analyzed studies, which take a relatively narrow view on crowdsourcing, this perspective enables a comprehensive depiction of task recommendation across the fragmented research landscape. The classification also provides a foundation to develop a comparative understanding of systematic differences regarding the design of task recommenders. In some cases, divergences can be attributed to fundamental characteristics of the respective system types, such as their organizational functions, the type of contributions they seek, and the design of their internal processes. Table A.1 lists the context of the analyzed studies grouped by system type. So far, no task recommendation research in the context of crowd rating systems has been identified.

Several studies are concerned with the development of recommendation approaches in the context of crowd processing systems. Ambati et al. (2011) and Yuen et al. (2011, 2012) propose recommendation engines for micro-task marketplaces such as Amazon Mechanical Turk. Organizations typically use these marketplaces to process large quantities of data, e.g., for classification, transcription, or search tasks. Taking a slightly different perspective, Satzger et al. (2011) describe an approach to preselect the crowd of potential contributors to such tasks based on their capabilities. To a similar end, Ho and Vaughan (2012) and Difallah et al. (2013) develop approaches for the semi-automated assignment of contributors to processing tasks. Although task assignment, as opposed to task recommendation, puts less emphasis on the self-selection principle, it is based on the same idea of matching tasks to suitable contributors.

A relatively large body of research focuses on task recommendation in the context of question answering communities such as Yahoo! Answers. While many questions in these communities represent individual rather than organizational problems, the function and the mechanics of the crowdsourcing systems implemented to solve them are similar. The heterogeneity of tasks approached in such systems presents an ideal use case for personalized recommenders, ranging from questions such as "What's a good travel destination in Europe?" to "How do I solve this Poisson distribution problem?" (Dror et al. 2011). The analyzed studies have developed approaches to match these questions with potential answerers. In a similar context, two studies address the problem of matching threads and potential experts in discussion forums (Castro-Herrera 2010; Zhou et al. 2009). Furthermore, Horowitz and Kamvar (2010) describe a social search engine that routes questions to persons that are most likely to answer them.

Four articles study the application of recommendation techniques in the context of crowd creation systems. Cosley et al. (2007) and Yuan et al. (2009) focus on the challenge of matching Wikipedia articles and editors. As the level of coverage and quality on Wikipedia grows, it gets increasingly difficult for potential editors to find suitable articles to contribute to. Jamjoom et al. (2009) present Cyano, a social networking enterprise application that uses a crowdsourcing approach to capture and maintain information on problem and change management processes for outsourced IT infrastructures. Among others, the application includes mechanisms to "recommend subject information that users may find interesting." Cosley et al. (2006) analyze different recommendation approaches to identify entries of MovieLens, an online movie database, that contributors might be willing to work on.

3.2.2 Recommender Techniques

Recommender systems research is an evolving field that explores an ever increasing variety of techniques to match users with items. While the exact algorithms used by these techniques change frequently, they all require certain kinds of knowledge in order to generate good recommendations. Based on a general distinction of the knowledge sources that are available to recommender systems (Burke and Ramezani 2011), the analyzed approaches were classified according to their use of individual, collaborative, and content knowledge sources. Table A.2 gives an overview of the results of this analysis.

3.2.2.1 Individual Contributor Knowledge

Knowledge about the individual users plays a central role in any personalized recommender system (Ricci et al. 2011). Burke and Ramezani (2011) distinguish between knowledge that is derived from a contributor's behavior and knowledge that is derived from their explicitly expressed opinions. In most cases, such knowledge is obtained by drawing on transactional information that reflects a contributor's past interactions with a system (Ricci et al. 2011), in particular with its tasks. As Horowitz and Kamvar (2010) put it, the rationale for this is that "if you want to be able to predict what kind of content a user will generate [...], first examine the content they have generated in the past."

Almost all of the identified approaches obtain knowledge from an individual's past contribution behavior, i.e., from the tasks that they have contributed to. Yuen et al. (2012) additionally include viewed and selected tasks. Some of the recommenders in question answering communities also consider the questions that an individual asked; however, this has no relevance for the context of organizational crowd solving systems with a clear separation of requesters and solvers. While most recommenders follow a passive approach in obtaining behavioral knowledge, two crowd processing systems include proactive mechanisms that probe a variety of contributor capabilities: Ho and Vaughan (2012) describe an up-front exploration phase during which tasks are assigned randomly to determine individual worker skill levels. Similarly, Satzger et al. (2011) propose to automatically assign assessment tasks with known results in order to improve the confidence levels for specific individual skills.

A small number of approaches derives additional knowledge from a contributor's opinions by considering flagged questions (Dror et al. 2011; Zhou et al. 2012), rated movies (Cosley et al. 2006), or refused recommendations (Horowitz and Kamvar 2010; Hu et al. 2008). In addition to opinions about specific tasks, a few recommender systems also rely on more generic background information. In three cases, contributors can specify lists of self-assessed skills or interests (Dror et al. 2011; Horowitz and Kamvar 2010; Satzger et al. 2011). Furthermore, two approaches extract information from structured and unstructured external sources, such as social network profiles (Difallah et al. 2013; Horowitz and Kamvar 2010).

As opposed to classic recommender systems, some task recommendation systems in crowdsourcing make use of an additional source of knowledge: the contributions provided by each individual. A large part of the analyzed approaches in crowd processing and crowd solving systems derives knowledge about a contributor's capabilities by assessing the quality of their individual contributions. In crowd processing systems, this assessment refers to the validity of a contribution, i.e., whether it was accepted by the requester or by an automated quality control mechanism. In crowd solving systems, contributions are assessed using a rating provided by the requester or other contributors. A number of crowd solving recommenders also incorporates the content of a contribution, i.e., the text of an answer, as an additional source of individual knowledge. One approach (Guo et al. 2008) even builds the contributor model entirely from an individual's answers without considering the question itself.

3.2.2.2 Content Knowledge

Content-based recommendation techniques aim to recommend items whose features match the information stored in a user's profile (Lops et al. 2011; Pazzani and Billsus 2007). Consequently, given the availability of knowledge about a contributor's interests and capabilities, the quality of content-based task recommendations depends on a recommender's ability to analyze and model the essential content of tasks.

The vast majority of the analyzed recommendation approaches make use of content knowledge. This knowledge is derived from unstructured textual descriptions of a task, such as its title or the instructions, and from structured classifications in form of keywords, tags, categories, or links. Further relevant task features include meta-data about the task, such as requester, reward, or allotted time.

As with individual knowledge, some task recommendation approaches in crowdsourcing derive additional content knowledge from the collective contributions to a task. Three recommenders in the context of crowd solving systems complement the given task description with the information contained in all (Dror et al. 2011; Zhou et al. 2009) or the best (Yan and Zhou 2012) answers to a question.

3.2.2.3 Collaborative Knowledge

Collaborative or social knowledge is knowledge about the larger community of users in a recommender system (Burke and Ramezani 2011). Unlike content-based recommendation approaches, pure collaborative filtering techniques do not require knowledge about the characteristics of items. Instead, they generate recommendations based on the interaction patterns of groups of users with the system (Su and Khoshgoftaar 2009). Collaborative filtering reflects the idea that a user's preference for an item is likely to be similar to that of other users with similar interaction patterns. Moreover, a user will have similar preferences for two items if other users have expressed similar preferences on them (Desrosiers and Karypis 2011).

All of the analyzed task recommendation approaches in crowd creation systems but few of the recommenders in other systems make use of collaborative knowledge sources. These approaches generate recommendations by considering patterns of collaborative behavior (Castro-Herrera 2010; Cosley et al. 2007; Jamjoom et al. 2009; Yuen et al. 2012), opinions (Cosley et al. 2006), or both (Dror et al. 2011). Four of the analyzed recommenders implement hybrid approaches that combine content and collaborative knowledge sources (Castro-Herrera 2010; Cosley et al. 2007; Dror et al. 2011; Yuen et al. 2012).

3.2.3 Recommender Evaluation

In order to find indications as to the general success and benefits of personalized task recommendation approaches, the analysis also comprised the evaluations conducted by the existing studies. A few studies had to be excluded as they either did not conduct an evaluation (e.g., Jamjoom et al. 2009) or the effect of task recommendation was not discernible (e.g., Horowitz and Kamvar 2010; Satzger et al. 2011). The remaining studies were examined with respect to the experimental settings, the respective samples, the evaluated properties and corresponding metrics, and any relevant results. Table A.3 gives an overview of the classification results.

Shani and Gunawardana (2011) describe three types of experimental settings to evaluate recommender systems in general: *Offline experiments* are simulated without user interaction on a pre-collected dataset. The vast majority of the analyzed task recommendation studies are evaluated in this manner. These evaluations are either based on publicly available datasets, such as from various question answering sites, forums, and Wikipedia, or on self-collected data, e.g., from dedicated tasks (Ho and Vaughan 2012; Yuen et al. 2011). In contrast, *user studies* assess the system through the interaction with a small group of test subjects, who receive actual recommendations in a controlled environment. Only one of the existing approaches (Difallah et al. 2013) is evaluated in a user study. *Online experiments*, finally, are performed by actually deploying the system in the field. Three studies in the context of crowd creation systems conduct such online experiments on Wikipedia (Cosley et al. 2007; Yuan et al. 2009) and the MovieLens platform (Cosley et al. 2006).

Various properties of a recommender system can be evaluated in these experimental settings (Shani and Gunawardana 2011). Most of the task recommendation studies, however, focus on assessing the *prediction accuracy* of their approaches. This is generally done in an offline experiment by calculating metrics that describe how well an algorithm can reproduce the recorded contributor-task relationships in a sample. The existing studies use a great variety of accuracy metrics, most of which are widely used in information retrieval, such as precision, recall, mean absolute error, mean average precision, mean reciprocal rank, or area under receiver operating characteristic (ROC) curve.

As these metrics are highly dependent on the respective (sub-)samples and are mostly used by the analyzed studies to compare different techniques, algorithms, and parameters, the isolated results shall not be discussed here. Instead, two studies that include a comparison with a random baseline are selected exemplarily: Yuen et al. (2011) show that their approach can predict contributor ratings with a mean average error of 0.5 compared to a value of 1.08 at random selection. Similarly, the approach described by Castro-Herrera (2010) significantly outperforms a random recommender at predicting if an individual has contributed to a specific forum thread. For the analyzed seven samples, the described content-based technique achieves a precision of 9.53–39.11 % and a recall of 14.29–49.04 %, while the random approach only achieves up to 2 % in both cases.

In addition to prediction accuracy, Hu et al. (2008) also assess the *coverage*, i.e., the fraction of all tasks that is recommended, whereas Zhou et al. (2009) assess the *scalability* of their respective approaches. As to the general benefit of task recommendation, however, perhaps the most interesting property of a recommender is its *utility*, i.e., "the value that either the system or the user gains from a recommendation" (Shani and Gunawardana 2011). All three online experiments and the user study calculate metrics that reflect the utility of the task recommender to the respective system by measuring the impact on contributions. Cosley et al. (2007) determine the fraction of recommendations that led to an actual contribution to Wikipedia as 4.3 % in contrast to a mere 1.0 % when giving random recommendations. In a longer-term study of the same recommender, Yuan et al. (2009) compare the mean levels of contributions between adopters (μ_a) and non-adopters (μ_{na}). While the adopters group on average contributed less before adoption ($\mu_a = 927 < \mu_{na} = 1050$), the contribution pattern was reversed after adopting task recommendations ($\mu_a = 789 > \mu_{na} = 563$). Difallah et al. (2013) measure utility as the fraction of valid contributions and find that their approach on average results in a 29 % improvement compared to traditional task selection on Mechanical Turk.

Cosley et al. (2006), finally, evaluate the impact of their task recommender on the contributions to Movielens, an online database that collects information on movies. They measure the number of unique editors, movies edited, and fields edited, which are, however, all dominated by a random baseline approach. This may be the result of a mismatch between the recommendation approach and the nature of the task. The applied collaborative filtering technique recommended relatively popular movies that had received high ratings by many people. Consequently, the movie information had probably been reviewed many times before with little room for improvement.

3.3 Discussion

This survey of the current state of the art shows that research on personalized task recommendation is yet in an early state and is spread across different types of crowdsourcing systems. The relatively small number of existing studies is an inherent characteristic of every literature review into an emerging topic (Webster and Watson 2002). Due to the variety of crowdsourcing systems and the fragmented nature of corresponding research, one cannot rule out the possibility that other relevant research is conducted in communities that remain disconnected from those analyzed. While an additional survey of crowdsourcing systems operating on the Web might reveal further instances of task recommendation approaches, it is problematic to obtain reliable information on their processes and efficiency without corresponding academic studies.

Table 3.1 Suggestions for future research on task recommendation

3.3.1 Applications and Limitations *Explore the potential of personalized task recommendation in application domains across all system types.*
3.3.2 Sources of Individual Knowledge *Improve contributor models by tapping into all available sources of individual knowledge, i.e., behavior, opinions, and contributions.*
3.3.3 The Role of Capabilities and Contributions *Analyze the unique characteristics of task recommenders regarding the consideration of capabilities in addition to interests and the derivation of knowledge from contributions.*
3.3.4 Connecting External Knowledge Sources *Investigate the integration of knowledge from external sources and the creation of meta-recommenders that exceed the boundaries of individual crowdsourcing systems.*
3.3.5 The Right Recommendation Technique *Draw on foundational knowledge from recommender systems research regarding the domain-specific selection, combination, and implementation of techniques.*
3.3.6 The Utility of Personalized Task Recommendation *Evaluate the actual utility of task recommendation approaches in large-scale online experiments to overcome the limitations of existing studies.*

Building on the analyzed studies, the following sections discuss a number of issues that future research should address in order to build a solid foundation for the systematic design of task recommenders. In line with Webster and Watson (2002), who state that an emerging topic will "benefit from exposure to potential theoretical foundations," the discussion draws on insights from mainstream recommender systems research where appropriate. Table 3.1 gives a brief summary of the discussed issues.

3.3.1 Applications and Limitations

Based on the early yet promising results of existing studies, personalized task recommendation appears to have the potential to enhance a variety of crowdsourcing approaches. Further research is required to determine the most promising applications of task recommenders within each of the system types and their limitations: Do the results from question answering communities hold for other instances of crowd solving systems? Is task recommendation in crowd creation systems useful for domains other than knowledge aggregation, such as journalism or software development?

While the literature search did not turn up any studies on task recommendation in crowd rating systems, this type of system may also hold significant potential for task recommenders. Many systems that seek to conduct collaborative evaluations, surveys, or predictions should be able to improve the overall quality of their results by recommending the respective rating tasks to contributors that are familiar with the particular items. Several indicators can be considered to determine suitable contributors, such as purchase or rental transactions for the rating of movies and travel bookings for the rating of hotels. Useful information could also be extracted from sources external to the particular crowd rating system. If, for instance, a contributor uses location-aware social networks to regularly check in at a restaurant, they probably possess the necessary knowledge to rate that particular place (see also Bulut et al. 2011). It needs to be considered, however, that in some situations task recommendation may also be detrimental to the outcome of crowd rating systems. When using crowd ratings to rank a set of items, for instance, a recommendation of individual items can easily cause a bias in the overall outcome. The same holds for surveys that aim to target a representative sample of individuals and not only those having a specific interest in the subject.

3.3.2 Sources of Individual Knowledge

The analysis of the literature has shown that personalized task recommendation systems can select from various sources of individual knowledge. Although in theory these sources are available to most crowdsourcing systems, the findings show that many existing approaches do not tap their full potential, which may partly be due to the lack of corresponding data in offline experiments. This section highlights a few issues toward the more structured creation of comprehensive contributor models.

Following Ricci et al. (2011), the relationship between a contributor and a task can be perceived as a series of interactions. Most evidently, interactions model a contributor's behavior, e.g., when a task is viewed, started, or completed. If possible, however, a system should also record interactions that reflect a contributor's explicitly expressed opinions about a task. Typically, these opinions involve a rating within scalar, binary, or unary categories (Desrosiers and Karypis 2011) and thus reflect varying degrees of individual preferences.

Sometimes a system cannot derive sufficient knowledge from a contributor's natural interactions with tasks. This is the case when a contributor only joined recently, which is referred to as the "new user" problem (Burke 2002), or if a system has limited interaction in general, e.g., if the available tasks are complex and time-consuming. In these situations, a recommender may be able to obtain more interaction knowledge by proactively asking contributors to consider a specific selection of tasks, such as generally well-liked or strongly varied ones. Recommender systems research provides further input on the application of such preference exploration or "active learning" techniques (Rubens et al. 2011).

Another option to "bootstrap" (Satzger et al. 2011) an individual contributor's profile is to ask contributors to provide task-agnostic background information on their general preferences. The entry of such information can be facilitated in various ways using structured forms, keyword or item lists, interview techniques, or gamification approaches (Hacker and von Ahn 2009). However, asking contributors to maintain explicit profiles imposes additional costs and may lead to an inaccurate depiction of their true preferences (Cosley et al. 2007; Morita and Shinoda 1994; Pazzani and Billsus 2007).

3.3.3 The Role of Capabilities and Contributions

The findings of this review confirm that personalized task recommendation differs from traditional recommendation scenarios in several ways. For one, the item suggestions generated by traditional recommender systems mainly reflect the intrinsic and extrinsic interests of an individual. In a crowdsourcing context, however, the usefulness of a task recommendation also depends significantly on the match with personal capabilities, such as expertise, skills, or domain knowledge. Following the self-selection principle (see Sect. 1.1), these capabilities are to a large extent reflected in the preferences that a task recommender can derive from a contributor's past behavior and opinions.

An additional, potentially more accurate, depiction of a contributor's capabilities may be obtained through an explicit assessment of their contributions. Depending on the organizational function of a crowdsourcing system, different assessments can be conducted. Systems that rely on heterogeneous contributions, i.e., crowd solving and crowd creation systems, can assess the absolute or comparative quality of individual contributions and, indirectly, the corresponding capabilities of their contributors. Indicators for such an assessment depend on the relevant contribution qualities as described in Sect. 2.2.2. Accordingly, they can range from automatically calculated objective measures to subjective ratings by the requester or other contributors. Systems with qualitatively identical homogeneous contributions, i.e., crowd processing and crowd rating systems, can instead assess the validity of a contribution and contextual factors such as the submission time. Again, different approaches are available, including automated checks, quality control mechanisms, or a manual inspection by the requester. Note that it is typically not possible to determine the truthfulness of contributions to crowd rating systems as they essentially represent subjective views (see Geiger et al. 2012).

Capability assessments bear particular relevance when the quality of a recommender system is judged not only from the perspective of its users but also from that of the organization that runs it. In many traditional scenarios, the satisfaction of the latter party follows directly from the satisfaction of the users' needs and wants. For instance, one of the most important functions of commercial recommender systems, increasing the number of sales, is achieved by pointing users to the items they are interested in and thus likely to buy (Ricci et al. 2011). In crowdsourcing systems, however, there are limits to this relationship. As opposed to a product recommender, the mere fact of accepting a task recommendation is not sufficient. A requester's goal is not simply to *sell* their tasks to anybody, but to obtain high-value contributions from individuals with the right capabilities. The crowd, after all, is not a crowd of passive consumers but rather one of service providers with diverse potential. Similar ideas have been brought forward in the context of the collaborative creation of value between producers and consumers (Kleemann et al. 2008; Zwass 2010). It is up to future research to determine to what extent task recommenders should consider the requester perspective and, in particular, balance diverging personal interests and capabilities when generating recommendations.

Contributions may not only be used as a source of individual knowledge, but they may also serve as a source of content knowledge to further carve out the features of tasks, e.g., in form of responses to crowd solving tasks or in form of changes to an existing Wikipedia article. Consequently, and in contrast to traditional recommendation scenarios, the items in a task recommendation system are determined not only by inherent or given features, but also through the collective input on these items. Some insight on this topic can be gained from recent recommender research on the value of user-generated content in social tagging systems (Hayes et al. 2007; Marinho et al. 2012; Niwa and Honiden 2006) and product reviews (Aciar et al. 2007).

In summary, personalized task recommendation approaches in crowdsourcing possess at least two unique characteristics that set them apart from most traditional recommendation scenarios: (i) they implicitly or explicitly model their users' capabilities in addition to their interests and (ii) they have the ability to draw on their users' contributions as an additional source of knowledge. The further study of these characteristic differences presents a valuable topic for joint research efforts that may eventually also add to the body of knowledge in mainstream recommender systems research.

3.3.4 Connecting External Knowledge Sources

Almost all of the contributor profiles that are built in existing task recommendation approaches are limited to information that is collected internally. Literature suggests that the quality of these profiles could be considerably enhanced by harnessing information that is external to the respective system. By perceiving a contributor not only as an isolated component of one particular system but as a connected individual that is part of many systems, a recommender can incorporate a rich variety of digital information already present on the Web, thus mitigating sparse data and creating more exact profiles. External information may come, for example, from other crowdsourcing systems (Demartini 2007), knowledge communities (Wang et al. 2013), social networks (Difallah et al. 2013; Horowitz and Kamvar 2010), social tagging systems (Markines et al. 2006; Niwa and Honiden 2006; Wu et al. 2006), blogs (Chua 2007; Kolari et al. 2008), expert directories (Becerra-Fernandez 2000; Hrastinski et al. 2012), and research publications (Buitelaar and Eigner 2008; Stankovic et al. 2011b). Some of these sources provide semantically linked data (Bizer et al. 2009) that enables the analysis of "user traces in structured form" (Stankovic et al. 2011b). Others are less structured and require the application of information retrieval and text mining techniques to extract relevant information (Balog et al. 2009; Deng et al. 2012; Gopal et al. 2011). While researchers have studied the use of these sources to find experts on specific topics, little seems to be known about integrating the information into unified, overarching user profiles, particularly in a crowdsourcing context.

Such integrated profiles also present a promising opportunity to create new task recommenders that exceed the boundaries of individual crowdsourcing systems and thus account for the scattered nature of the crowdsourcing landscape. Most of the identified recommenders have been developed in the context of crowdsourcing systems with a large amount of tasks, such as marketplaces for crowd processing and solving or extensive crowd creation platforms. Naturally, these systems have access to a large user base and can collect a rich amount of information. Smaller or less known systems, however, often struggle to attract a sufficient number of suitable contributors. Although some of the tasks in these systems might be a perfect match to the interests and capabilities of a particular individual engaged in crowdsourcing activities, chances are high that this person will never discover the respective system. The development of meta-recommenders, which maintain integrated contributor profiles and recommend tasks from multiple systems, could mitigate this problem and increase the visibility of suitable tasks.

On a cautionary note, an aggregation of personal data and organizational tasks from multiple sources will raise a number of privacy and legal concerns. Nevertheless, such integrated approaches present a promising avenue to increase the efficiency of crowdsourcing by bringing about a shift from the current platform-centric to a more content-oriented perception of crowdsourcing tasks.

3.3.5 *The Right Recommendation Technique*

Mainstream recommender systems research provides an extensive body of knowledge on all stages of recommender development. Based on this analysis of the crowdsourcing literature, however, it appears that the design of personalized task recommendation systems could benefit from a more substantial application of such "foundational knowledge" (Hevner et al. 2004). By adapting this knowledge to a crowdsourcing context, researchers can draw on insights regarding, for instance, the selection and implementation of suitable techniques.

All of the surveyed task recommendation approaches are based on variations of content-based or collaborative filtering techniques. Few approaches make use of hybrid recommendation techniques, although recommender systems literature explicitly highlights the increase in efficiency when combining the benefits of different techniques and thus mitigating their respective weaknesses (Burke 2007). Furthermore, few approaches consider additional information such as demographics, context, or social relations, which have been used in other types of recommender systems to improve the quality of recommendations. Community-based or social techniques (Ricci et al. 2011), for instance, use measures of social relatedness to generate recommendations from the ratings of related users. Seen by some as a subtype of collaborative filtering, these techniques also do not require knowledge about the items themselves.

As with any recommender system, the choice of the employed approach should be made with the characteristics of the application domain in mind. Burke and Ramezani (2011) discuss several factors that impact the availability and quality of different knowledge sources and, consequently, how they determine the application of corresponding recommendation techniques or combinations of techniques. See Sect. 4.3.2 for an example of how such considerations can inform the fundamental design of a recommender system.

In addition to guiding the choice of the recommendation approach, recommender systems research also provides insights on further common design issues, such as dealing with sparse data (Desrosiers and Karypis 2011; Kim et al. 2011), various biases (Piramuthu et al. 2012), or malicious users (Burke et al. 2011; Chung et al. 2013). Moreover, when it comes to actually implementing a recommendation technique, existing libraries already provide a variety of basic algorithms. Among the most popular open source toolkits are, for instance, Apache Mahout,[1] MyMediaLite,[2] LensKit,[3] GraphLab,[4] and Easyrec.[5]

3.3.6 The Utility of Personalized Task Recommendation

It appears that one of the most fundamental issues for future research on personalized task recommendation is the empirical evaluation of its general utility to the crowdsourcing process. Existing research has presented promising results in the context of crowd creation systems; the majority of studies, however, focuses on comparing the accuracy of various recommendation techniques in offline experiments. While these experiments demonstrate that, in theory, predicting contributor preferences works well, they have a number of shortcomings that limit their generalizability.

Firstly, the available datasets usually lack knowledge about a contributor's preferences for most of the tasks. An offline evaluation therefore cannot reliably assess the accuracy of a prediction for a contributor-task pair that is not contained in the sample. Moreover, the recommender's ability to build a comprehensive profile of individual contributors, and thus its potential to make good recommendations, is limited by the sources of knowledge present in the available samples.

Secondly, in many cases, a large part of the data that underlies an offline experiment reflects observed contributor behavior. As discussed in Sect. 1.1, such

[1] http://mahout.apache.org/

[2] http://www.mymedialite.net/

[3] http://lenskit.org/

[4] http://graphlab.org/

[5] http://easyrec.org/

implicit behavior may already be suboptimal due to a minimization of search costs, the inadequacy of sorting criteria (if any), or manipulation by some requesters, e.g., through frequent reposting of tasks (Chilton et al. 2010). Consequently, assessing a recommender's accuracy in terms of how well it reproduces such datasets cannot convey any knowledge about its ability to improve the match between contributors and tasks.

Thirdly, as Shani and Gunawardana (2011) put it, "accurate predictions are crucial but insufficient to deploy a good recommendation engine." Even if a task recommender succeeds at making predictions that accurately reflect contributor interests and capabilities, it is unknown if users will accept them, nor if they actually lead to improved results: Do recommendations, on an overall scale, have an impact on the efficiency and the outcome of a crowdsourcing system?

In order to address these issues, future research will need to conduct field studies in the context of real crowdsourcing environments. Adequate metrics have to be developed to determine the utility of task recommendation approaches for both requesters and contributors. In addition, recent research (Herlocker et al. 2004; Shani and Gunawardana 2011) highlights the importance of further recommender system properties, such as novelty, serendipity, or diversity, which might be of particular interest in a crowdsourcing context.

Chapter 4
Design of a Third-Party Task Recommendation Service

Building on the insights that were gained during the literature review, this chapter describes the prototypical design of a third-party task recommendation service called *metacrowd*, which integrates with existing crowdsourcing systems that do not feature built-in recommendation mechanisms. In line with the identified need for large-scale field research, metacrowd provides a platform for studies in real-world, productive contexts. The setup permits complete control over the recommender system design and, consequently, facilitates the extensive collection of related research data and the systematic study of specific design aspects.

In contexts in which personalized task recommendation proves a useful approach, metacrowd also has practical applications. As laid out in Sect. 1.2, the enhancement of existing crowdsourcing systems with recommendation mechanisms has the potential to improve the contributor satisfaction as well as the overall efficiency and the results of a system. Eventually, metacrowd could also be used to realize integrated contributor profiles and cross-platform meta-recommendations as described in Sect. 3.3.4.

The next section briefly outlines the most essential requirements. Section 4.2 then summarizes the architectural foundations of the developed prototype. This is followed by a detailed description of the individual system elements in Sects. 4.3–4.6. The chapter concludes with a discussion of the current design and future opportunities for improvement in Sect. 4.7.

4.1 Requirements

The basic functionality of the metacrowd prototype is to generate personalized task recommendations based on a contributor's interests and capabilities. To this intent, metacrowd needs to maintain profiles of the individual contributors that use the service (see Sect. 3.3.2) and match these profiles with the available tasks. Metacrowd is conceptualized as a third-party recommender system and, as such,

D. Geiger, *Personalized Task Recommendation in Crowdsourcing Systems*,
Progress in IS, DOI 10.1007/978-3-319-22291-2_4

must implement mechanisms to obtain the required data on contributors and tasks from existing crowdsourcing environments.

Metacrowd is based on a modular design that supports various crowdsourcing platforms, which allows its use in different system contexts and also provides a foundation for cross-platform recommendations. The prototype features an implementation for the Mechanical Turk platform, which was described in Sect. 1.1. Due to its large task volume and contributor numbers, Mechanical Turk is a well-suited starting point for field studies on personalized task recommendation. The modularity principle also facilitates future research of a comparative nature, such as experiments with different recommendation techniques.

Despite its prototypical nature, metacrowd will be deployed in a productive environment in order to conduct an online evaluation with real-world data. Consequently, it needs to support the stable, scalable, and unsupervised use by unknown clients. Task recommendations should incorporate the latest data on the available tasks to enable the instant recommendation of new tasks and avoid the recommendation of outdated tasks. This is challenging due to the high throughput of the Mechanical Turk platform. In order to provide a dynamic user experience, the recommender also needs to promptly incorporate interactions with the crowdsourcing platform or the metacrowd system. This includes, perhaps most importantly, a user's feedback on recent recommendations. Finally, the application needs to account for the sensitive nature of the stored contributor data and allow for a high level of user anonymity.

4.2 Architecture

The metacrowd prototype is built on a client-server architecture that comprises a central backend and a browser extension that runs on every client machine. The backend implements the business logic of the application and aggregates all data required by the recommender system. It is based on a combination of the *domain-driven design* (DDD) approach and the *Ports and Adapters* architectural style, which are briefly explained in this section. The various parts of the backend, which reflect this architecture, are described in detail in Sects. 4.3 through 4.5.

The browser extension serves two main purposes. Firstly, it provides a user interface (UI) to the application, whose central element is a sidebar that is added to supported crowdsourcing websites. Contributors can use this sidebar, for example, to register with the system, receive recommendations for interesting tasks, and interact with these recommendations. As opposed to other UI options, such as a separate website or a browser-independent client application (e.g., a Java client), the use of an extension enables metacrowd to directly integrate into the familiar environment of the crowdsourcing platform, enabling a seamless user experience.

Secondly, the extension collects information about the interactions between contributors and crowdsourcing platforms, such as the tasks that a contributor participates in. In combination with other sources of knowledge, this information is used to build profiles of contributor interests and capabilities and thus serves as input to the recommender system (see Sect. 4.3.2). Due to the contributor-specific nature of the corresponding data, it can typically, as in the case of Mechanical Turk, not be accessed from the outside. Consequently, the use of an extension that runs within the contributor's browser environment is the only feasible way to record the relevant interactions. The extension is described in detail in Sect. 4.6.

4.2.1 Domain-Driven Design

Domain-driven design (Evans 2004; Vernon 2013) is an iterative approach to software engineering that focuses on developing a model of the business logic of the underlying real-world problem domain. This model is only described in terms of the concepts and the language of the business domain and is therefore understood by all stakeholders, including both technical and domain experts. From an implementation perspective, the domain model is directly represented by a code layer that sits at the heart of the software application. DDD provides a set of techniques and principles to guide the development of domain models, delineate the boundaries of different subdomains, and describe the integration between multiple domains.

Domain-driven design also defines a number of "tactical design patterns" that describe the building blocks of a domain model, some of which are depicted in Fig. 4.1. *Entities*, for instance, model domain objects that are characterized by a unique identity and their own lifecycle, such as a person. *Value objects*, on the other

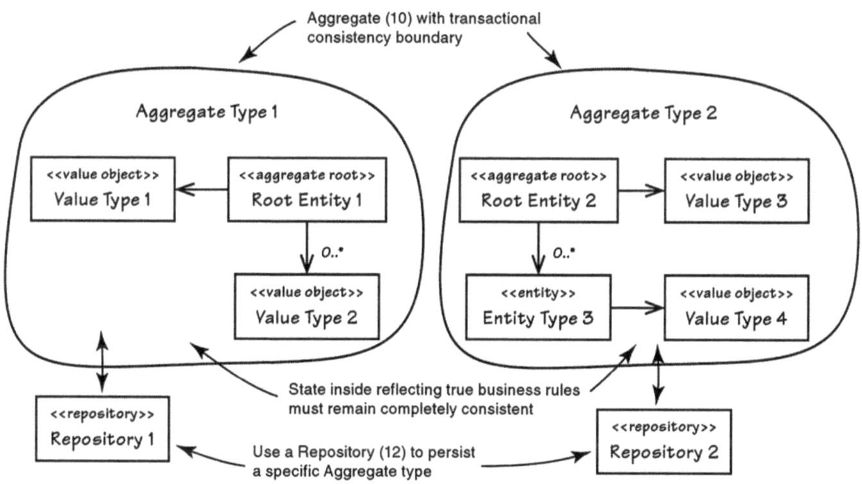

Fig. 4.1 Tactical design patterns in domain-driven design (Vernon, Vaughn, Implementing Domain-Driven Design, 1st, ©2013. Printed and electronically reproduced by permission of Pearson Education, Inc., New York, New York)

hand, do not have a unique identity but instead are defined by the sum of their values, such as an address. DDD emphasizes that entities and value objects are not merely modeled as a set of attributes but also in terms of the associated behavior. They are thus different from the simple data objects used by other modeling approaches, which are often called "anemic" (Fowler 2003) due to their lack of functionality. An *aggregate* describes either a single entity or a collection of entities and value objects that form a unit of transactional consistency with a single point of access, a root entity. Aggregates encapsulate domain invariants, i.e., business rules that must always be strongly consistent. Any process that spans multiple aggregates, consequently, is not expected to ensure transactional consistency, but instead allows for an acceptable delay in reaching an eventually consistent state. The aggregate design pattern thus provides a very good fit with some of the characteristics of non-relational data stores, as will be described in Sect. 4.5.3.

In order to facilitate eventual consistency across aggregates, *domain events* may capture important occurrences, such as state changes, that are of interest to other domain objects, which can then react on the event in an asynchronous manner. The creation of aggregates can be delegated to *factory* classes or methods, while their storage and retrieval is handled by *repositories*, which abstract the domain model from the actual storage infrastructure. *Domain services* model functionality that does not naturally represent an operation of any particular entity or value object. A *module*, finally, describes a named and highly cohesive set of domain objects with low coupling to other modules. This corresponds to the package language construct in Java.

4.2.2 Ports and Adapters

While the domain model describes an application's business logic, the Ports and Adapters architecture (Cockburn 2005) focuses on the relationship of an application with its environment. Ports and Adapters aims to address a common cause of design problems: the "entanglement between the business logic and the interaction with external entities." This issue presents itself, for instance, in the leakage of business logic into the user interface or in the close coupling of business logic to technical infrastructure concerns. Multiple architectures with similar intents have (re)surfaced in recent years, such as the "clean architecture" (Martin 2012) or the "onion architecture" (Palermo 2008).

The Ports and Adapters architecture lays out a well-defined boundary around the inside of an application, which is depicted in Fig. 4.2. The functional specification of the application, as expressed in use cases or user stories, is made at this boundary and is not concerned with the characteristics of external actors, such as the number and type of clients or output mechanisms. Instead, all communication with the environment is abstracted by a set of conceptual *ports* in the application boundary. Ports define the API of an application and reflect all of the functionality that the application provides to or requires from the environment. Every port serves a

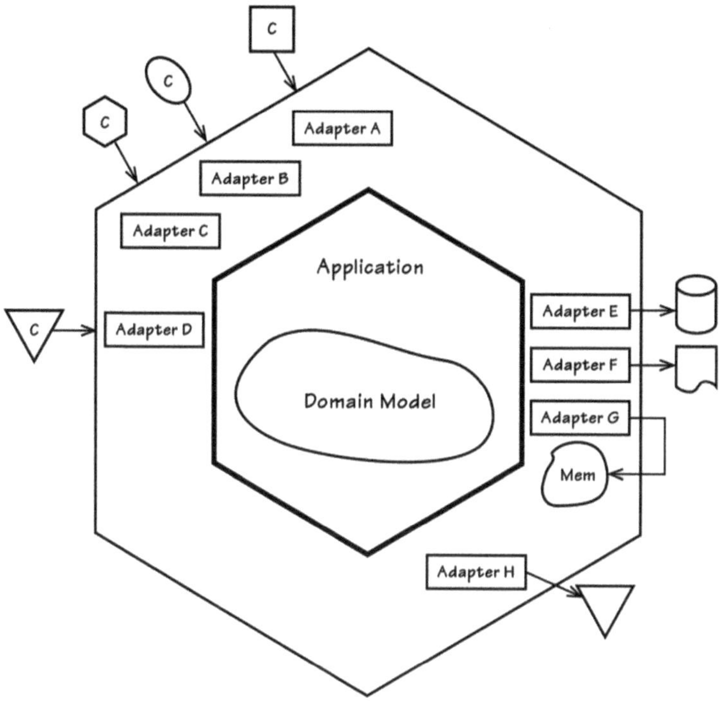

Fig. 4.2 Ports and Adapters (Vernon, Vaughn, Implementing Domain-Driven Design, 1st, ©2013. Printed and electronically reproduced by permission of Pearson Education, Inc., New York, New York)

specific conversational purpose and can have several *adapters*, which convert the respective API definition into the format needed by a particular external entity and thus plug this entity into the application. In line with the notion of primary and secondary (or supporting) actors in use case descriptions (Cockburn 2000, p. 53 ff.), two types of ports and adapters can be distinguished.

Primary adapters, depicted on the left side of Fig. 4.2, connect entities that drive the application, which means that these entities initiate the conversation. Primary adapters may provide various application interfaces to handle the requests of both human and computational clients. From an implementation perspective, when using a domain-driven design approach, primary adapters typically plug into the application by invoking the methods published by a thin set of services around the domain model. These *application services* reflect the primary ports of an application and act as the single client to the domain model. Application services do not contain any business logic of their own and orchestrate the functionality of aggregates, factories, repositories, and domain services in a secure and, if required, transactionally consistent manner.

Secondary adapters, depicted on the right side of Fig. 4.2, connect entities that are driven by the application, which means that the conversation with these entities is initiated by the application. Typical secondary adapters connect different storage mechanisms and specific system services or act as clients to other systems. Following the Dependency Inversion Principle (Martin 1996), which states that low-level details always depend on high-level abstractions, dependencies in a Ports and Adapters architecture always point 'inwards' and the application is never concerned with the specific adapters. Consequently, when applying the DDD patterns, the repositories and domain services that represent the secondary ports of an application are designed as software interfaces. Secondary adapters are implementations of these interfaces. A particular implementation is typically plugged into the application by using either a service locator or a dependency injection design pattern (Fowler 2004).

Besides the commonly found primary "user-side API" and secondary "data-side API" (Cockburn 2005), many applications feature additional conversational ports, for instance, to provide administration interfaces or to synchronize data with other systems (see Sect. 4.5.2). In order to emphasize this fact, the Ports and Adapters architecture is usually depicted by use of a hexagonal layout. In contrast to a one-dimensional layered drawing, the edges illustrate the potential existence of more than two ports, which may each be connected through multiple adapters. For this reason, and although the number six is of no particular importance in this context, the architecture is also known as the Hexagonal architecture.

The main benefit of the Ports and Adapters architecture, as opposed to, e.g., a traditional Layers architecture (Buschmann et al. 1996), is its ability to design and run an application completely independently of any particular external actors. The resulting applications are highly flexible, as adapters for different interfaces and technologies may be added or removed at any time. This also enables developers to primarily focus on the application and defer low-level framework and technology decisions, thus leading to more informed choices. Equally important, all parts of a Ports and Adapters application are easily testable, for example, by connecting automated regression test suites to its primary ports or through the use of in-memory mock databases on the secondary ports.

4.3 Domain Model

The metacrowd application, at its core, models the domain of personalized task recommendations in a crowdsourcing context with a particular focus on the Mechanical Turk platform. This core model comprises four modules that reflect the basic domain concepts involved in a task recommendation system. Section 4.3.1 first introduces the essential characteristics of the contributor, task, and interaction modules. The recommender module will then be described in detail in Sect. 4.3.2. Following the presentation of the core domain, Sect. 4.3.3 briefly outlines how the application models user identity.

4.3.1 Contributors, Tasks, and Interactions

Contributors, tasks, and interactions are fundamental concepts of any crowdsourcing domain. As such, they exist independently of the recommender system, but are vital to its operation. Metacrowd models each of these concepts, their respective attributes, and their behavior as a single entity that comprises a set of value objects. A particular contributor, task, or interaction entity and all related value objects form a transactionally consistent aggregate, which is persisted and retrieved through the use of corresponding repositories.

4.3.1.1 Contributors

A contributor models a person that potentially submits contributions to the tasks available on a specific crowdsourcing platform. Contributors are uniquely linked to a user of the metacrowd application. While the prototype focuses on the Mechanical Turk platform, a user may, from a conceptual perspective, be linked to multiple contributors, one for each supported platform. The integration of an individual's platform-specific profiles increases the recommender's knowledge base and may therefore enable improved and cross-platform recommendations, as described in Sect. 3.3.4.

Contributors on the Mechanical Turk platform are not only characterized in terms of their interactions with tasks, but also by the qualifications they have obtained. Qualifications are measured on various types of scales. Many qualifications are assigned or calculated automatically by the platform, others can be requested by the contributors themselves and may require passing a corresponding test. Some examples for qualifications on Mechanical Turk include 'HIT approval rate,' 'location,' 'adult content qualification,' or 'German language test.'

4.3.1.2 Tasks

A task model, in its most basic manifestation, contains at least a title, a description, and a status that is either open or closed. Open tasks can currently receive contributions from interested individuals; they are thus eligible for recommendation. Closed tasks, on the other hand, are not open for participation anymore and should not be recommended. Once closed, however, a task will still be available from the repository since the stored information is needed for recommendation generation.

Metacrowd's Mechanical Turk module extends this basic task model to account for several platform-specific characteristics. Tasks on Mechanical Turk can be described by their requesters in a semi-structured manner using keywords, such as 'survey,' 'transcribe,' or 'research.' In addition, since Mechanical Turk is a paid crowdsourcing platform, contributions to a task are paid with a specified reward upon successful completion. The task model also includes the id and name of the requester and the number of available assignments.

Moreover, requesters on Mechanical Turk can restrict their tasks to a subset of the worker population by setting formal qualification constraints, which declare a condition on the qualifications that potential contributors need to possess. Such constraints are defined using a comparison operator that depends on the qualification's type (greater than, equals, not less than, less than, has been granted), and a corresponding value. Contributors qualify for a given task only if their qualification values fulfill all of the declared constraints.

4.3.1.3 Interactions

An interaction describes a specific relationship between a contributor and a task as outlined in Sect. 3.3.2. Metacrowd models different types of interactions, all of them including a date and additional type-specific information. If an interaction of a particular type occurs repeatedly, the respective repositories only store the latest information for a given contributor-task pair.

The metacrowd prototype models two types of relationships that reflect the direct interaction between a contributor and a task: submissions and ratings. A *submission* interaction, meaning that a contributor submitted a contribution to a task at a specific point in time, records a behavioral information that is available on all crowdsourcing platforms. A *rating* interaction, by contrast, records an explicitly expressed positive or negative preference that reflects the opinion of the respective contributor.

Since the Mechanical Turk platform per se does not offer its users the possibility to rate tasks, this feature is introduced by the browser extension described in Sect. 4.6. The prototype employs a ten-point rating scale that is familiar from other recommender systems, such as on the Internet Movie Database.[1] Note that the merits of different rating scales are a subject of ongoing research in the recommender systems community (Cosley et al. 2003).

Two further types of interactions model a contributor's reaction to a task when it was recommended. If contributors are interested in a particular recommendation, they can *examine* the corresponding task further by clicking on the recommendation. On the other hand, contributors can also explicitly *reject* recommendations if the corresponding task is not relevant to them. In both cases, the interactions contain the date of the user reaction and additional information on the context of the recommendation, which is used for evaluation purposes.

[1] http://www.imdb.com/

4.3.2 *Recommendation Generation*

Recommendation generation in the metacrowd prototype is handled by a dedicated domain service, which has been optimized for the Mechanical Turk environment. All recommendations are generated live and are personalized to the respective contributor. This section describes the rationale for the fundamental recommendation technique, followed by a detailed description of the implemented algorithm.

4.3.2.1 Selecting a Recommendation Technique

As discussed in Sect. 3.3.5, the choice of a recommender system's essential recommendation technique should be driven by the availability and quality of knowledge sources and therefore depends heavily on the nature of the application domain. One factor that bears a particular relevance to the recommendation of crowdsourcing tasks is the "churn" of the item space (Burke and Ramezani 2011). Many crowd processing systems, in particular marketplaces such as Mechanical Turk, represent a recommendation domain that is defined by high churn, meaning that most tasks are short-lived and new tasks arrive frequently. Each task can only receive a finite number of contributions, corresponding to the number of assignments. While some tasks may comprise hundreds or thousands of assignments, the large majority completes within a few hours to a few days at most (Ipeirotis 2010).

High churn domains pose a challenge for classic collaborative filtering techniques, which depend on the availability of transactional data to determine overlaps in preference patterns. As new tasks enter the system, they have not been involved in any interactions; a situation which is often referred to as the "new item" (Burke 2002) problem. Unless the number of recommender system users is comparatively large, collaborative techniques will often not be able to record the interactions required to generate confident recommendations before a task becomes unavailable. An efficient personalized task recommendation for the Mechanical Turk domain therefore needs to make use of content-based techniques, which generalize from individual tasks by deriving characteristic features that also transfer to new tasks. For a discussion of complementary similarity measures and hybrid recommendation approaches, see Sect. 4.7.

The quality of content-based recommendations depends heavily on a system's ability to model user preferences in terms of the available item features. Following the reasoning in Sect. 1.1, the prototype recommender focuses on features that reflect the type or topic of a task. Firstly, the recommender incorporates the requester-specified keywords of a task. While the Mechanical Turk platform comprises a wide variety of tasks, the vocabulary of their keywords remains fairly constant and thus enables a good characterization of contributor preferences. Secondly, the recommender uses the identity of the requester itself as a more implicit proxy for the type or topic of a task. Based on evidence from the task data collected during metacrowd's development, tasks of a given requester are often very similar to each other. Moreover, these tasks resemble each other in qualitative characteristics that transcend the actual work assignment, such as the adequacy and promptness of payment. It is thus reasonable to assume that contributors who preferred tasks of a particular requester might be interested in other tasks of that requester.

4.3.2.2 The Recommender Algorithm

Most personalized recommender systems are based on a model of (positive and negative) user preferences, a set of candidate items, i.e., all items eligible for recommendation, and one or several similarity metrics (Ricci et al. 2011). Using the similarity metric, the recommender aims to match user preferences with the available items in order to predict their usefulness or suitability to the user. Based on these predictions the recommender system then generates a ranked list of items, in this case tasks, to be suggested to the user.

The metacrowd recommender derives contributor preferences from the set of tasks that a contributor has interacted with in the past. This comprises all tasks that a contributor has submitted or rated as well as those tasks that were rejected during earlier recommendations. As suggested in Sect. 3.3.2, this approach derives preferences from both contributor behavior, as reflected in the task submissions, and contributor opinions, as expressed in form of task ratings and rejected recommendations. All of this information is available from the contributor-task interactions that are stored in the corresponding repositories (see Sect. 4.3.1). In order to account for changes in preferences over time, as well as for performance reasons, the recommender only considers the 300 most recent interactions of each type.

The recommender also models the strength and the direction of the preference. If available, the respective preference value is determined directly as the rating that a contributor has given to a task. In the case of Mechanical Turk, this is as a value between 1, representing an extremely negative preference, and 10, representing an extremely positive preference. Where no rating is available, the recommender maps the recorded interactions to predefined values on the scale. In order not to overemphasize the impact of preferences inferred from single interactions, the recommender assigns a moderately positive preference value, such as 7, to submitted tasks and a moderately negative preference value, such as 3, to rejected recommendations.

The contributor-specific set of candidate tasks is determined as a subset of all currently open tasks available in the repository. In order to avoid the recommendation of already known tasks, the recommender removes all tasks that the particular contributor has previously interacted with. In addition to submitted, rated, and rejected tasks, which are used to derive contributor preferences, the recommender also ignores tasks that the contributor has already examined upon a previous recommendation. Users of the metacrowd service can further narrow down the set of candidate tasks on a per-recommendation basis by setting two filter parameters for the minimum number of available task assignments and for the minimum reward per contribution; see Sect. 4.6.1 for details. Finally, the remaining set of tasks is restricted to those candidates whose qualification constraints are fulfilled by the contributor's qualification values.

At this point, the recommender applies a similarity measure to match the model of contributor preferences with the available candidate tasks. Based on the considerations in the previous section, the metacrowd prototype implements a content-based task similarity metric using keywords and requesters. This metric does not depend on the input of a large number of users—as would be the case with collaborative filtering techniques—and allows the recommendation of new tasks as soon as they enter the system.

The recommender calculates a similarity value for each pair of tasks from the set C of candidate tasks and the set P of tasks for which a preference value has been derived. The similarity of a pair of tasks ($c \in C$, $p \in P$) is a number between 0 and 1 and is defined as an equally weighted combination of its requester and keyword similarity:

$$\mathrm{sim}(c, p) = \frac{1}{2} \cdot \mathrm{sim}_{\mathrm{requester}}(c, p) + \frac{1}{2} \cdot \mathrm{sim}_{\mathrm{keyword}}(c, p).$$

The requester similarity simply reflects if two tasks have the same requester as a binary value of 0 or 1, whereas the keyword similarity quantifies an overlap in their keywords as a decimal value between 0 and 1. For two sets of keywords, the similarity value is determined as their Jaccard coefficient (also called the Tanimoto coefficient), which is defined as the size of their intersection divided by the size of their union. If both sets of keywords are empty, the similarity is 0. Formally, the keyword similarity of two tasks with non-empty sets of keywords K_c and K_p is defined as

$$\mathrm{sim}_{\mathrm{keyword}}(c, p) = \frac{|K_c \cap K_p|}{|K_c \cup K_p|}.$$

Using the combined requester and keyword similarities, the recommender determines a preference estimate for each candidate as the average of the past preference values for all similar tasks, weighted by the respective similarity value. Formally, the preference estimate for any $c \in C$ is given by

$$\mathrm{pref}_{\mathrm{estimated}}(c) = \frac{\sum_{p \in P} (\mathrm{sim}(c, p) \cdot \mathrm{pref}(p))}{\sum_{p \in P} \mathrm{sim}(c, p)}.$$

The resulting preference estimates are used to sort all candidates in descending order. If two tasks have an identical preference estimate, the estimate that is based on more data points, i.e., on a higher number of non-zero similarities, is preferred. In order to ensure a minimum level of recommendation quality, the recommender only considers candidates with a preference estimate above a defined threshold, such as the center of the preference scale, and with a non-zero similarity to at least two preferred tasks.

In a final step, the sorted list of candidates is filtered to account for the fact that some requesters regularly enter batches of near-identical tasks within a short time frame. In order to avoid redundant recommendations, candidates are filtered if they share the same requester and the same set of keywords and if their titles are very similar. Title similarity is determined using the Levenshtein distance, a string metric that belongs to the family of edit distances (Levenshtein 1966). The Levenshtein distance denotes how many individual character modifications (deletions, insertions, or substitutions) are required at a minimum to transform one string into another. Based on samples of real task sets, the recommender prototype was configured to consider two titles similar if their edit distance is less than 30 % of the longer title's length. When removing redundant candidates, the recommender keeps the one with the most assignments.

Filtering ends as soon as the requested number of non-redundant recommendations (ten by default) has been determined or if no more candidates remain. The recommender then returns these candidates as a list of recommendation value objects. These objects include the essential attributes of the recommended task, such as title, description, requester, reward, and number of assignments. They also include additional information about the performed calculations, such as the preference estimate or the number of data points, which is used for evaluation purposes.

The described algorithm is summarized in Fig. 4.3. The implementation is based on the widely-used Apache Mahout framework (Owen et al. 2012). A number of optimizations, such as a layered caching approach, have been applied to minimize

- Load two distinct, contributor-specific sets of tasks:
  - the set of (positively or negatively) preferred tasks
    - all tasks submitted, rated, or rejected in the (limited) past
  - the set of candidate tasks
    - all non-preferred, non-examined open tasks, filtered by qualifications and optional user criteria
- Calculate a preference estimate for each candidate task:
  - Determine the content-based similarities to all preferred tasks as an equally-weighted combination of the identity of requesters (0/1) and the ratio of shared keywords to the union of keywords (Jaccard coefficient).
  - Estimate the preference as the average of the preferences for all similar tasks, weighted by the respective similarity value.
- Recommend the top candidate tasks:
  - Exclude candidates that are redundant, based on less than two similarities, or below a defined preference threshold.
  - Return the requested number of recommendations with the highest estimates.

Fig. 4.3 Metacrowd's content-based recommendation algorithm

the cost and time required for a recommendation generation cycle. These optimizations are described in detail in Appendix B. As a consequence, the average recommendation times during the prototype evaluation consistently stayed below one second.

4.3.3 *Identity Management*

In addition to the functionality provided by the core domain, the metacrowd application implements a generic identity management subdomain (Vernon 2013, p. 52) that allows for the unique identification and authentication of users. The lightweight solution comprises a user entity, a corresponding repository, an encryption service, and an authentication service.

Every user has a unique username, a password, a unique email address, and methods to change these properties. Specifying an email address is optional to allow users to remain anonymous. Upon the creation of a new user or upon a password change, a given plaintext password is run through an encryption service, which creates a hash value that is stored with the user. The current encryption service implements a standard Blowfish password hashing scheme (Provos and Mazières 1999).

The authentication service authenticates a user identified by a username and password combination. Authentication is performed by retrieving the user entity for the given username from the repository and asking the encryption service to validate the given plaintext password against the stored hash value. Upon successful validation, a security credential that identifies the authenticated user is created and returned.

4.4 Application Services

The prototype contains a set of thin application services that reflect use case groupings with respect to the five domain modules. As described in Sect. 4.2.2, these services act as the single client to the domain model and expose the application's functionality to the primary adapters of the system in a secure and consistent manner. In addition to basic user management and access to the recommender, metacrowd provides interfaces to feed the necessary platform-specific data on contributors, interactions, and tasks into the application.

The identity application service provides methods to register a new user as well as to retrieve and update the information of the currently authenticated one. Additionally, it provides a login method that invokes the authentication service and, upon successful authentication, returns the created security credential. This security credential is required by all application services that handle user-specific data (with the exception of the login and registration methods) to clearly identify the authenticated user. In this manner, it also acts as an implicit authorization check ensuring that users are only allowed to operate on their own data.

The contributor application service handles information that characterizes platform-specific contributor entities. Among others, the service provides a method to update a contributor's qualification values. It also enables all other application services that perform contributor-related operations, i.e., the interaction and recommender one, to determine the contributor entity for the authenticated user and a given crowdsourcing platform. Upon a user's first recorded interaction with a particular platform, the service creates and stores a new contributor entity.

The task application service provides methods to add or update tasks and to retrieve all open tasks, which are exclusively used by the task synchronization adapter described in Sect. 4.5.2. Additionally, it provides a query method that retrieves tasks in a given time period by title and requester. This functionality is used in the process of synchronizing a contributor's submission history, as described in Sect. 4.6.2.

The interaction application service manages direct interactions between pairs of contributors and tasks. To this intent, it provides methods to record new submissions and ratings as well as methods to retrieve existing ratings, which can then be displayed to the end user. Additionally, the service allows the retrieval of a contributor's submitted tasks within a specific period of time. This functionality is also required during the synchronization of submission histories.

The recommender application service, finally, enables the interaction with the recommender system. As such, it offers two kinds of functionality. Firstly, it allows clients to request a list of platform-specific, personalized recommendations with an optional specification of the number of recommendations and any filter criteria. Secondly, the service provides methods to record a contributor's feedback on a recommendation by either examining or rejecting it. Internally, this information is stored as an interaction with the respective task (see Sect. 4.3.1).

4.5 Adapters

Various primary and secondary adapters connect the metacrowd backend application with the environment. In particular, the system includes an API to communicate with the browser extension, which is described in Sect. 4.5.1, and a platform-specific synchronization of relevant crowdsourcing tasks, which is described in Sect. 4.5.2. Several adapters connect the application to the infrastructure services provided by the Google App Engine platform; those are laid out in Sect. 4.5.3. Furthermore, the prototype contains a number of adapters that complement its core functionality, such as performance profiling interceptors, dependency injection modules, or an administration dashboard, which provides usage statistics and maintenance capabilities.

4.5.1 Extension API

The browser extension communicates with the metacrowd backend through a Web service API that corresponds to the representational state transfer (REST) architectural style (Fielding 2000). The primary adapters that provide this API translate the functionality exposed by the application services into a set of resources and resource collections. A *resource* is an abstract concept for an object that is accessible by a uniform resource identifier (URI). Clients can operate on a resource only through regular HTTP methods such as GET, PUT, POST, and DELETE. Any data that is exchanged between client and server uses a standard Internet media type (Freed and Borenstein 1996), in this case the JavaScript Object Notation (JSON), specified in Bray (2014).

In order to retrieve or update information, for instance, on the rating of a task with a particular id, the browser extension sends a GET or PUT request to the URI associated with the corresponding backend resource, located at '/api/interactions/{taskId}/rating'. Similarly, the extension can retrieve a list of recommendations by sending a GET request to the URI of the recommendation resource collection at '/api/recommendations'. The API provides access to the functionality of all application services with the exception of the user-independent task application service, which is only accessed by the synchronization adapter described in the next section.

Identity management is handled by two separate resources, which invoke the corresponding methods of the identity application service. The user resource allows clients to register a new user account (POST), to retrieve (GET) and update (PUT) the account information, or to remove (DELETE) the account. Once registered, clients can log in and out using the authentication resource. Upon a successful login, the security credential that is returned by the identity application service is stored in an HTTP session. The corresponding session id is typically stored in a cookie on the client side and is transferred with every request. As described in Sect. 4.3.3, almost all application services that handle user-specific data require this security credential to identify the authenticated user. The corresponding resource methods are therefore mapped to an additional authentication filter, which intercepts all requests, attempts to retrieve the security credential from the session, and passes it into the resource method. If no security credential is available, the filter denies access by returning an HTTP 401 'Unauthorized' status.

The extension adapters are built using the RESTEasy[2] framework, an implementation of the Java API for RESTful Web Services (Pericas-Geertsen and Potociar 2013). JAX-RS enables the definition of resources as annotated Java classes whose methods implement the supported HTTP operations. An additional plugin for the Jackson[3] data-processing suite takes care of the automatic parsing and generation of JSON code, which allows the convenient use of simple Java data transfer objects as method arguments and return values.

---

[2] http://resteasy.jboss.org/

[3] http://github.com/FasterXML/jackson

4.5.2 Task Synchronization

In order to generate valid recommendations, metacrowd requires current information on the tasks available for contribution. While the most straightforward way would be to query the respective crowdsourcing platform directly, Amazon Mechanical Turk does not provide a public API to access this information in a programmatic way. Instead, metacrowd obtains the necessary task data through a third-party platform. The Mturk-Tracker (Ipeirotis 2010) calculates continuous statistics on the available tasks and their various characteristics, such as types, rewards, and requesters. In order to collect this data, the platform performs regular crawls of the complete list of HIT groups on the Mechanical Turk website using a screen scraping approach that converts the HTML code into structured information, which is then stored into a database. Every crawl takes about ten minutes to iterate through all available tasks, continuously updating Mturk-Tracker's database along the way. When a crawl is done, a new one starts from the beginning of the task list.

The Mturk-Tracker provides a RESTful API[4] that consists of two resources. The main 'hitgroupcontent' resource can be used to obtain data on every task that the crawler ever encountered. This comprises static properties, such as title, description, requester, reward, keywords, and qualification constraints. It also includes the latest value of mutable properties, such as the number of available assignments and the 'last update,' which refers to the last time that a task was encountered by the crawler. The 'hitgroupstatus' resource provides additional information on every individual update, most notably the start date of the corresponding crawl.

Metacrowd's task synchronization adapter is invoked every minute by a cron job and implements a two-step process to incrementally synchronize the local task repository with data obtained from the Mturk-Tracker API. In a first step, the adapter retrieves all tasks that were updated since the last synchronization from the 'hitgroupcontent' resource. A dedicated task translator processes the JSON response received from the API and creates corresponding task entities. Among others, the translator converts any HTML-encoded characters in the task titles into standard Unicode and transforms the plain text qualification constraints (e.g., 'HIT approval rate is not less than 85') into proper value objects with a qualification name ('HIT approval rate'), a constraint type constant ('NOT_LESS_THAN'), and a value (85). Once the entire API response has been processed, all tasks are marked as open and are stored in the local repository using the task application service, overwriting any older versions. Finally, the time of the most recent update is stored as a milestone for the next invocation of the adapter.

In a second step, the adapter determines outdated tasks that are marked as open in the local repository but are not available anymore on the Mechanical Turk platform. Since the Mturk-Tracker does not explicitly state if a task is still available, the adapter needs to deduce this information from the last update. To this intent, the adapter retrieves and stores the start date of the most recent Mturk-Tracker crawl

[4] http://mturk-tracker.com/documentation/mturk-tracker/api/

from the 'hitgroupstatus' resource. If this date differs from the one stored during the previous synchronization, a new crawl has started and the previous crawl is assumed to have finished, which means that it has recorded the complete list of available HIT groups within a particular time frame. Consequently, the adapter considers all tasks that have not been updated since the start date of this complete crawl as outdated and closes them.

Note that the detection of outdated tasks has an inherent delay, which may cause the system to temporarily recommend tasks that are not available anymore: In the worst case scenario, a task is completed immediately after it has been encountered at the beginning of a crawl. It is only after the subsequent crawl has finished, i.e., up to two crawl spans (about 20 minutes) later, that the task can be marked as closed. In practice, however, this delay only affects tasks with very few remaining assigments.

In addition to the regular synchronization, the task adapter enables the on-demand retrieval of tasks. Every time the application records a new interaction for a given task id, it checks whether a corresponding task entity is available in the local repository. If it is not, the task adapter tries to obtain the task from the Mturk-Tracker API. This occurs if the task is new and has not yet been synchronized or if the scheduled synchronization of open tasks was interrupted for an extended period of time. It can also mean that the task was missed by the MTurk-Tracker crawls entirely, in which case the interaction will be stored anyway but metacrowd will not have any content data to be used during recommendation generation.

4.5.3 App Engine Infrastructure

The metacrowd backend is written in Java and runs on the Google App Engine, a scalable Platform-as-a-Service offering for Web applications in the Google cloud (Google 2014a). Applications that are deployed on the App Engine execute within one or several isolated virtual machines called instances. Incoming client requests are automatically distributed to a particular instance with free capacities. By scaling up the number of instances, an App Engine application is able to serve any number of clients. App Engine includes a scheduler engine that can be configured to start or shut down instances depending on the current and predicted load.

App Engine offers a large number of infrastructure services to its applications. Several secondary adapters have been implemented to enable access to the corresponding functionalities. This section describes those services that are essential for the operation of the prototype and how they are connected to the application.

4.5.3.1 Persistent Storage

Most notably, the App Engine platform provides access to Google's non-relational ('NoSQL') High-Replication Datastore, a schemaless object data store that holds objects known as entities (Google 2014b). Every entity is identified by a given kind and a unique key and can have several properties. In contrast to relational databases, whose tables are typically bound to a single machine so the various integrity constraints can be efficiently enforced, the Datastore distributes and replicates entities across a large number of geographically distributed nodes. This architecture allows the Datastore to provide high availability and scalable performance.

Besides retrieving entities by their key, the Datastore supports queries on the kind and properties of entities, which are served by pre-built indexes. In contrast to relational databases, the time required to execute a query scales with the size of the result set and not with the total number of entities. In combination with the distributed nature of data storage, this makes the Datastore ideally suited for very large amounts of data, such as those in a recommender system. In order to implement this performance guarantee, the Datastore does not support certain types of queries, such as join operations or inequality filtering on multiple properties. In many cases, these limitations can be addressed by a carefully reasoned entity design. If necessary, the unsupported operations can also be implemented in application code, which, however, negates the performance advantages of non-relational databases.

As opposed to many other non-relational databases, the Datastore supports atomic transactions that may span operations involving up to five entity groups. An entity group is a unit of transactional consistency that comprises a root entity and, optionally, a number of child entities that are permanently dependent on the parent. Entity groups are thus conceptually equivalent to the aggregate design pattern promoted in the domain-driven design approach (see Sect. 4.2.1). In order to efficiently maintain the consistency constraint, all entities that belong to a particular group are stored on the same Datastore node. As a consequence, the rate of changes that can be performed on a single entity group is limited to currently about 1–5 write operations per second (Google 2014c).

In general, due to the time required to synchronize data across nodes and to update the indexes that are used to serve queries on entity properties, non-relational databases are required to make a trade-off between high scalability and data consistency. The Datastore always provides strong consistency for the retrieval of an entity by its key and for queries within a given entity group. Global queries, i.e., queries that span multiple entity groups, however, only provide eventual consistency, which means that changes to an entity may not be immediately reflected by subsequent queries. This fact needs to be accounted for in the application design and highlights the importance of well-designed aggregate boundaries.

The metacrowd application connects to the Datastore through adapters that implement the repository interfaces for the user, task, contributor, and interaction aggregates defined in the domain model. Internally, the adapters make use of Objectify,[5] a convenience interface specifically designed for the Datastore, which exposes all of its non-relational characteristics. Objectify provides several

[5] http://code.google.com/p/objectify-appengine/

abstractions from Google's low-level APIs, but remains transparent enough to allow applications to benefit from the full capabilities of the Datastore.

Objectify transforms instances of the plain Java classes that represent the domain model entities into the entity objects stored by the Datastore and vice versa. To control this transformation, Objectify provides annotations that designate the field holding an entity's unique id, which is stored as part of its Datastore key, and annotations that define indexes for any fields that are to be used in queries. Once the entity classes have been properly annotated, the Objectify API can be used to store corresponding instances in the Datastore, retrieve them by their id, or execute queries on the indexed fields. Objectify supports all native Datastore features including batch processing, asynchronous operations, and transaction handling. Moreover, it provides mechanisms to define polymorphic entity hierarchies, implement lifecycle callbacks, and facilitate logical schema migrations, among others.

4.5.3.2 Distributed Caching

App Engine also features a distributed, in-memory data cache that is available to all instances of an application. The Memcache (Google 2014d) is approximately 5–10 times faster than the persistent Datastore and may thus considerably speed up operations on frequently accessed data (Google 2013). It stores entries using a key-value model and is available in a free version, which may evict entries at any time, and a paid version, which reserves a dedicated amount of space for an application. As discussed in Appendix B, the use of caching to optimize speed and costs is a crucial element in optimizing the recommender design.

The Objectify data access API used by the repository implementations integrates consistent Memcache support, which allows the transparent use of this distributed cache as semi-durable storage. Every read, save, or delete operation that is performed on Datastore entities through Objectify will also update the corresponding cached values. By using Objectify for all Datastore calls, metacrowd's adapters thus ensure that the cached entities stay up-to-date at all times.

4.5.3.3 Attribute Validation

In addition to the functionality provided by the Datastore and Objectify, metacrowd's repository adapters implement support for the constraint-based validation of attribute values. Among the attributes that require validation are those that depend on direct user input, such as usernames, passwords, and email addresses. Validation is also used to detect inconsistencies during data synchronization with other systems (see Sect. 4.5.2) as early as possible and thus avoid consecutive faults. The adapters make use of Java's Bean Validation framework (Bernard 2013) to define value constraints directly on the respective attributes and validate them immediately before an entity is saved. If a constraint has been violated, the adapters throw a corresponding exception, which is eventually raised with the user or stored in the application logs.

Besides checking isolated value constraints, some attributes also need to be validated for uniqueness across all entities of a certain kind. This is required, for example, to ensure that no two users have the same username or email address and that a user entity is related to at most one contributor entity per supported crowdsourcing platform. Whereas the Datastore enforces the uniqueness of entity keys, it currently, as opposed to relational databases, provides no native support for unique non-key properties. This capability, however, was implemented as part of the metacrowd prototype in form of a unique property service, which is described in Appendix C.

4.5.3.4 Asynchronous Processing

Some of metacrowd's use cases rely on asynchronous processing capabilities. Typically, this involves processes that take a significant amount of time to complete but do not provide an execution result that is of immediate relevance to their callers. Invoking these methods asynchronously avoids unnecessary waiting times on the client side. Furthermore, App Engine limits the execution time of synchronous client requests to 60 seconds. This time frame is sometimes insufficient to complete processes that are heavily based on the communication with external entities or that potentially require several retries, e.g., due to transaction collisions. An example of such a process is the synchronization of submission history entries described in Sect. 4.6.2.

Due to restrictions that apply to automatically scaled instances, App Engine does not allow the execution of standard Java background threads. Instead, it provides an API to schedule units of work called *tasks* in a scalable queue (Google 2014e). Depending on the queue configuration, scheduled tasks will be processed either automatically according to a predefined rate (push queue) or through on-demand consumption (pull queue). In contrast to synchronous requests, a task may take up to ten minutes to execute and, unless otherwise specified, is automatically scheduled to retry if it fails to complete. Tasks can also be added to the queue in the context of a transaction, in which case they are guaranteed to execute if and only if the transaction successfully commits. Each task is characterized by an endpoint URL and an optional payload of parameter values. As an alternative to setting up HTTP request handlers, e.g., Java Servlets, for each distinct task, the queue API also provides an interface for 'deferred tasks' that consist of a single method with a set of arguments and are processed by a default handler.

Metacrowd includes an infrastructure adapter that facilitates the asynchronous invocation of methods via the task queue. Most notably, the implementation comprises a method annotation and an associated interceptor that converts the otherwise synchronous invocation of an annotated method into a deferred task that corresponds to the described interface. Once the task has been scheduled in a predefined queue, the control flow returns to the method caller. In order to set up an asynchronous invocation, it suffices to annotate the invoked method without making any changes to the method caller.

4.6 Browser Extension

Metacrowd's central backend connects to individual client side environments by way of a browser extension that uses the API described in Sect. 4.5.1. The browser extension is available for Chrome and Firefox, which are currently the two most popular browsers, in particular among Mechanical Turk workers (see Sect. 5.2.3). The extension primarily contains a *content script* that is injected into the web pages of supported crowdsourcing platforms. Content scripts are JavaScript files that are executed in an isolated environment (Google 2014f; Mozilla 2014). This allows them to access and modify the document object model (DOM) of the respective page but keeps them separate from any other JavaScript code on the page.

Content scripts that are packaged as a dedicated extension have several advantages when compared to plain *user scripts*, which are simple JavaScript files that are run by third-party extensions such as Greasemonkey[6] in Firefox or Tampermonkey[7] in Chrome. A browser extension is essentially a software component that adheres to the standards defined by a particular browser. It is subject to the browser's security framework and contains a *manifest* file that, among others, explicitly specifies the required permissions, such as access to user data on particular pages (e.g., mturk.com) or to cross-domain resources (e.g., at the metacrowd backend). Extensions further enhance user trust by being hosted in the official browser vendor stores, i.e., the Chrome Web Store[8] and the Mozilla Add-Ons[9] platform. These stores include detailed descriptions, screenshots, user ratings, and, in the case of Firefox, involve an in-depth review by experienced community members. Moreover, as with any software component, extensions can bundle and reuse existing JavaScript libraries in order to deal with specific implementation concerns.

The content script is built on top of the popular Backbone[10] library, which facilitates the design of client-side user interfaces by separating data management and presentation logic (Osmani 2012, 2013). A Backbone application defines a set of models and collections of models that encapsulate and manage the data associated with different parts of the application. Models can be synchronized with backend resources via a set of save, retrieve, and delete operations. Backbone provides built-in support to translate these operations into asynchronous HTTP requests that are sent to the respective resource URI of a RESTful Web service API,

[6] http://addons.mozilla.org/firefox/addon/greasemonkey/

[7] http://tampermonkey.net/

[8] http://chrome.google.com/webstore/category/extensions

[9] http://addons.mozilla.org/firefox/

[10] http://backbonejs.org/

such as the one described in Sect. 4.5.1. A Backbone application also defines a number of views, which are JavaScript objects that build and maintain the DOM elements that users can interact with. Views typically render a visual representation of models by using a templating mechanism that merges the model data with HTML code. Implementing an observer pattern, views register with models to receive change events and consequently update themselves automatically. In addition to these basic building blocks, the content script makes use of the Marionette[11] library, which complements Backbone with a set of best-practice design patterns and commonly used supporting functionality.

The various models, collections, views, and other parts of the content script have been written as individual JavaScript module files based on the CommonJS[12] specification. All modules have been designed to be largely independent of the particular browser environment. In order to package the extension, a programmatic build system called Gulp[13] is set up to perform a multi-step process. It first merges all modules, including the corresponding HTML templates, into a single script file using the Browserify[14] tool. Additionally, it obtains any required third-party libraries from the Bower[15] repository. The script files are then combined with additional resources, such as preprocessed style sheets,[16] images, fonts, and the browser-specific manifests. Finally, the entire folder structure is compressed into component archives for Chrome and Firefox, which can be uploaded to the extension stores using the respective developer interfaces. Figure 4.4 shows a screenshot of the metacrowd extension as listed in the Chrome Web Store.

Once a user has installed the extension from the store, the content script is injected into all web pages whose URL matches a predefined pattern that corresponds to the supported crowdsourcing platforms, such as 'http[s]://*.mturk.com/*'. Every time the content script is injected, it performs a number of basic setup tasks. This includes adding containers for the UI elements to the loaded page, initializing the connection with the backend, and loading any modules that are specific to the currently displayed crowdsourcing platform. As soon as the script has been set up, it passes control to a workflow manager, which displays and coordinates the various views of the user interface as described in the following section. Moreover, upon a successful authentication of the metacrowd user, the workflow manager triggers several contributor data synchronization processes, which will be described in Sect. 4.6.2.

[11] http://marionettejs.com/
[12] http://wiki.commonjs.org/
[13] http://gulpjs.com/
[14] http://browserify.org/
[15] http://bower.io/
[16] http://sass-lang.com/

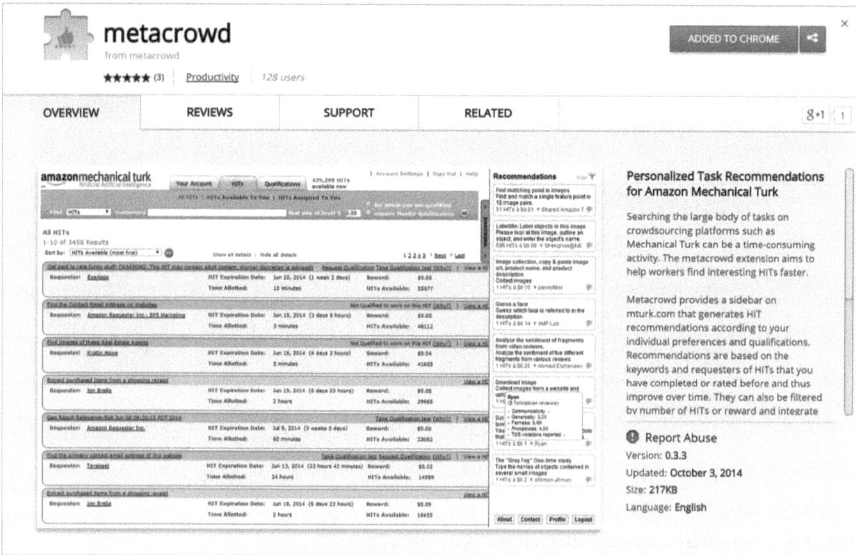

Fig. 4.4 The metacrowd extension in the Chrome Web Store

4.6.1 User Interface

Metacrowd's main user interface consists of a sidebar that is displayed to the right of the supported crowdsourcing platforms, as depicted in Fig. 4.5. The sidebar includes a small metacrowd label that is always visible and can be clicked to hide the sidebar, e.g., if more space is required to work on a task, and to show it again. The visibility state of the sidebar is persisted in local browser storage and restored at every invocation. A series of buttons at the bottom of the sidebar provide access to supporting functionalities, such as a contact form or an about page, which will be displayed in a separate window-style box. The about page, which is also displayed the first time the extension is used, contains a brief description of how metacrowd works and instructions on its usage. It also lays out a privacy policy, as depicted in Fig. 4.6, that lists all types of information that the extension collects and to what purpose this is done.

When it is passed control, the workflow manager performs a brief authentication check with the authentication resource in the backend. If the backend cannot retrieve a security credential from the user's HTTP session, authentication fails and the sidebar presents a login form. An additional link allows users to register for a new account. Registration requires the user to specify a username, a password, and an optional email address; all of which can be changed at a later time in the sidebar's user profile section. In order to register, users must further declare to have read and accepted metacrowd's privacy policy.

Once a user has been successfully authenticated, the extension accesses the backend's recommendation resource collection, thus triggering the live

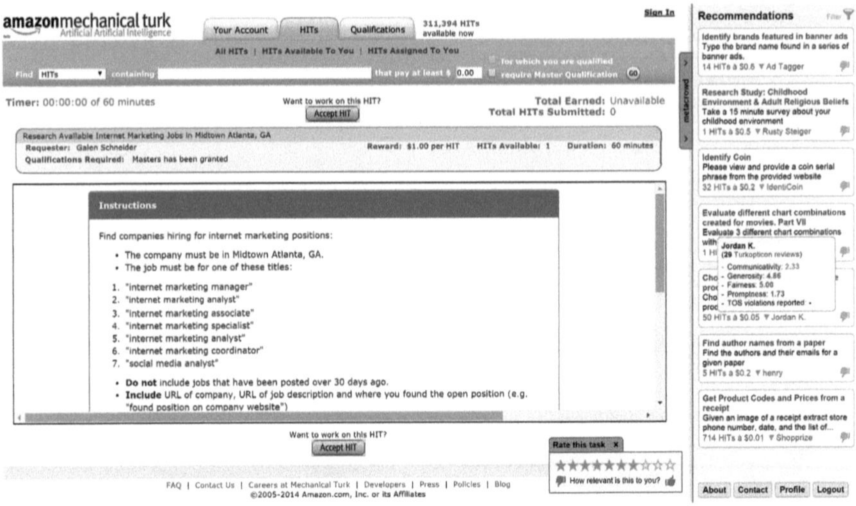

Fig. 4.5 User interface of the metacrowd browser extension

About metacrowd ×

Privacy Policy

In order to generate personalized recommendations, metacrowd needs to learn about your preferences and qualifications. This information is kept strictly confidential and will only be used to determine matching tasks and, in anonymized form, for academic research purposes.

The following is an exhaustive list of the information that is stored:

- Your metacrowd username, password (hashed), and email address (optional) as provided during registration
- Your ratings on selected tasks
- The ids of the tasks that you submit and have submitted in the past days (as listed in your platform profile) and the corresponding submission time
- Your qualifications (as listed in your platform profile)
- Your feedback to specific recommendations
- A session id and a corresponding cookie to keep you logged in
- Anonymous metacrowd usage data to evaluate and improve the application

In particular, metacrowd does **not** store:

- Third party user credentials, e.g., your login data on crowdsourcing platforms
- The content of task submissions
- Interactions with websites other than mturk.com

Fig. 4.6 Metacrowd's privacy policy

recommendation generation process described in Sect. 4.3.2. By default, the extension retrieves the top ten recommendations among all qualified tasks that the user has not interacted with before and displays them in the sidebar. A filter icon on the top of the recommendations view allows the user to further restrict the candidate tasks by specifying a minimum number of task assignments and a minimum reward per contribution. Note that the latter does not represent the hourly reward, since Mechanical Turk does not provide any information on the time required to work on a task. The filter option does, however, allow contributors to easily restrict recommendations to paid tasks only or, assuming fair payment, to distinguish between quick, small tasks and those that require more effort. Similar to the visibility state of the sidebar, the filter criteria are persisted in local browser storage.

Every recommended task is depicted as a small box that aggregates the essential information. This includes the task title and description, the number of available assignments, and the reward per contribution. The description is truncated after two lines, which is indicated by an ellipsis symbol, and can be shown in full using a mouse-over tooltip. Additionally, the task boxes display the name of the task requester and, if available, the corresponding Turkopticon[17] ratings. Turkopticon (Irani and Silberman 2013) is a third-party platform that enables Mechanical Turk workers to review requesters with respect to several criteria such as generosity, fairness, processing time, and compliance with Amazon's terms of service. As soon as the extension has retrieved a list of recommendations from the metacrowd backend, it sends an asynchronous request with the corresponding requester ids to a Web service provided by Turkopticon. The returned requester ratings and a link to the detailed reviews are shown in a small overlay box that is available through an icon on the bottom of the respective recommendation.

Users can interact with a particular recommendation in two ways (see Sect. 4.3.1). In order to further examine a recommended task, a user can click on its title in the sidebar. The extension notifies the backend of this interaction and then loads the respective page of the task on the crowdsourcing platform, which provides more detailed information and allows the user to contribute to the task. Note that the page change triggers a reinjection of the extension's content script, which will retrieve an updated list of recommendations. Alternatively, a user can reject a recommendation by clicking on a thumbs-down icon next to the task box. The extension transmits this information to the backend and removes the task from the list of recommendations.

In addition to the interaction with recommendations, metacrowd introduces a rating functionality that allows users to express their opinions on any task, not just the recommended ones. Rating tasks, as stated in metacrowd's instructions, is an important means to improve personal recommendation quality. The rating functionality integrates directly with the individual task pages of the Mechanical Turk platform. To this end, the extension performs a pattern matching approach on the loaded document URL and, if it finds that the current page displays a single task,

[17] http://turkopticon.ucsd.edu/

shows a rating box on the bottom of the page. The box asks the user to rate the personal relevance of this task on a ten-star scale. When the user selects a star, the rating is sent to the backend from where it will also be retrieved the next time the rating box for this task is displayed.

4.6.2 Synchronizing Contributor Data

Besides the information that is transferred as part of a user's interaction with the metacrowd system, the recommender relies heavily on information about the inherent interactions between a contributor and the supported crowdsourcing platforms. As described in Sect. 4.2, such contributor-specific data is not publicly available and can only be accessed through the client-side extension. In the case of Mechanical Turk, the collected data comprises the contributor's qualification values and the tasks in which a contributor participates.

4.6.2.1 Qualification Values

In order to only recommend tasks to which a user is allowed to contribute, metacrowd needs to obtain information on the qualifications that a contributor has gained. These qualifications are listed in a dedicated section of the contributor's profile on the Mechanical Turk website. While the metacrowd backend could potentially obtain this data directly from the website by logging in as the individual contributors, this would require storing the respective Amazon passwords in clear text and thus violate common privacy guidelines.

Instead, metacrowd employs a client-side synchronization approach using the browser extension. Every time the extension is loaded, i.e., at every user interaction with the Mechanical Turk website, it polls the backend for synchronization instructions. In regular intervals, such as more than 15 minutes after the last successful synchronization, the browser extension is instructed to resynchronize the contributor's qualification values. Upon receiving such a command, the extension initiates a screen scraping approach by sending asynchronous HTTP requests for the corresponding pages of the Mechanical Turk profile. As long as the contributor is logged into the Mechanical Turk website, these requests are processed within the scope of the active session.

Using CSS selectors, the extension extracts a list of qualification name and value pairs from the relevant parts of the received HTML code. This data is subsequently transmitted to the backend's contributor resource, which triggers an update of the respective contributor entity, overwriting any old qualification values. Finally, the current time is stored to determine the next synchronization cycle.

4.6.2.2 Task Submissions

The metacrowd browser extension implements a live tracking mechanism that detects submissions to tasks as they are made. At every user interaction with the Mechanical Turk website, the extension matches the current document URL with a predefined submission pattern. Whenever a user has submitted a contribution, the respective task id is extracted from the URL and transferred to the backend, which records a new submission interaction.

While the tracking approach ensures the prompt availability of submission information in the backend, it has two shortcomings. Firstly, it is entirely dependent on the permanent operation of both extension and backend components: If a user logs out, uses a different browser, or upon temporary backend failure, metacrowd will not be able to track any submissions. Secondly, the approach suffers from a variation of the common "new user" problem (Burke 2002): Although most users that register with the system may have been active on the Mechanical Turk platform for a long time, the metacrowd recommender initially has no knowledge on their preferences. This is the equivalent to the new item problem described in Sect. 4.3.2.

In order to address these shortcomings, metacrowd implements a complementary synchronization of past submissions based on a contributor's profile on the Mechanical Turk website. The profile contains a section that lists the title and requester id of all tasks that were submitted within the past 45 days, aggregated by submission date. Similar to the synchronization of qualification values described in the previous section, the browser extension regularly polls the backend for instructions. If the last complete synchronization dates back more than one day, the extension is instructed to crawl the contributor's submission history using an asynchronous screen scraping process. This approach enables users that have previously submitted a number of contributions to the Mechanical Turk platform to receive personalized recommendations almost instantly once they start using metacrowd.

Depending on the number of entries to be synchronized and on the client's connection speed, crawling the submission history may take from a few seconds up to several minutes, in particular during the first synchronization. Consequently, the synchronization process is performed incrementally for each date in the submission history. Starting at the first date that has not yet been synchronized, the extension extracts all entries and transmits them to the backend, which stores the corresponding submission date as a synchronization milestone. If the synchronization process is interrupted before the entire list of submission dates has been processed, e.g., because the contributor interacts with the Mechanical Turk website in between, the process will pick up from this date. Once the entire list of submission dates has been crawled, the synchronization is marked as complete.

Upon receipt of a batch of submission history entries for a particular date, the backend triggers an asynchronous background job that compares these entries to the contributor's submission interactions already stored in the repository for that day. Since the list of submissions on the Mechanical Turk website does not specify the corresponding task ids, tasks need to be compared by title and requester id. If the comparison shows that a submission history entry has not yet been recorded, the system attempts to determine the id of any task that was open within the relevant submission period and matches the given title and requester id. Note that if the respective requester has published multiple tasks with the same title in a short time frame, this approach may select a task that is not identical but very similar to that of the original submission. Once a task id has been identified, a new submission interaction is stored.

4.7 Discussion

The implemented prototype provides a fully operational third-party task recommendation service. Through the use of platform-specific synchronization mechanisms, metacrowd presents an approach to design task recommendation mechanisms around existing crowdsourcing systems. While the prototype focuses on the Mechanical Turk platform, the modular design allows future versions to add support for other systems. Besides the practical applications of enhancing crowdsourcing systems with task recommendation mechanisms, metacrowd provides a versatile platform to conduct field studies within the respective environments. The next chapter proceeds to evaluate the described design with these purposes in mind.

From a conceptual perspective, several avenues could be taken to further improve upon the prototype's recommendation quality. As one option, metacrowd's models of contributor preferences could be extended with more information. This may include considering additional types of interactions with the platform. Moreover, the impact of a particular interaction on the recommendation calculation could be weighted continuously according to its age. In contrast to the currently implemented hard cut at 300 interactions per type, this might allow the recommender to produce a smoother and more realistic model of changes in preferences over time.

Metacrowd's contributor models could also be enriched with further information on individual capabilities. While the prototype models the capabilities reflected implicitly in a contributor's past behavior and opinions as well as those explicitly formalized in qualifications, an additional source of knowledge may be available in the explicit assessments of contributions (see Sect. 3.3.3). Mechanical Turk includes such assessments in form of 'accept' or 'reject' decisions made by the task requester for every received contribution. Other platforms might feature more fine-grained scales to assess contribution quality.

Furthermore, external sources of knowledge could prove a valuable addition to the data that is collected within a particular crowdsourcing system. As outlined in Sect. 3.3.4, various sources, such as social networks or expert directories, may potentially provide further insights into a contributor's interests and capabilities.

This also includes the idea of connecting user-specific data across crowdsourcing systems to create more comprehensive, integrated contributor models.

Another direction that future versions of metacrowd may pursue is to refine the similarity metric that is used to match contributor profiles and candidate tasks. Most evidently, it might be possible to incorporate further relevant task features into the content-based metric. On a more fundamental level, a promising alternative to calculating item similarities could be a classifier-based approach that directly learns user-specific weights for individual feature characteristics. In some cases, such an approach might provide a better identification of those characteristic values that have proven to be a particularly good predictor of user preferences in the past, such as specific keywords or requesters.

On the other hand, the current approach could easily integrate collaborative information to determine the similarities of items. Assuming an increasing number of metacrowd users and thus a higher probability of overlaps in the recorded preference patterns, it might be worth exploring the use of collaborative filtering techniques, which by themselves do not require any knowledge about task features. As discussed in Sect. 4.3.2, this should be particularly effective in crowdsourcing systems with a lower churn rate. Still, the Mechanical Turk recommender might implement a hybrid approach (Burke 2007) that combines a content-based similarity metric for new tasks with a collaborative one for tasks that have been involved in a sufficient number of interactions.

Once again, these considerations show that the recommendation technique for a particular metacrowd module will need to be tailored to the characteristics of the respective crowdsourcing domain. Moreover, recommender systems research highlights that the merits of a specific implementation can ultimately only be determined by deploying the system in the target environment (see Sect. 3.3.6). Using the metacrowd platform, future research can set up corresponding experiments that comparatively evaluate a wide variety of recommender configurations.

Chapter 5
Personalized Task Recommendation in the Field

This chapter evaluates the potential of the developed prototype to realize and study personalized task recommendation on existing crowdsourcing platforms and thereby improve the match between contributors and available tasks. To this intent, the following sections present a sequence of three studies that were conducted along the process of field testing and deploying the metacrowd service on the Mechanical Turk platform. Section 5.1 first describes a pilot study that gathered initial feedback from a small group of experienced contributors. Section 5.2 then lays out the details of a large-scale survey on general task search behavior and on the perceived utility of the metacrowd service. Section 5.3, finally, analyzes a dataset that was gathered during an extended period of productive use and discusses the challenges identified during the evaluation.

5.1 Pilot Study

In order to obtain qualitative user input on metacrowd's design, early versions of the service were put up for review in two of the leading forums[1] for Mechanical Turk workers. The announcement included a brief introduction to the service, links to the browser extensions on the Chrome Web Store and the Mozilla Add-Ons platform, as well as a request to express feedback, concerns, and suggestions for further improvements. Figure 5.1 shows an exemplary excerpt.

The responses to this announcement revealed the need for several extensions to the initial version of the prototype. In particular, it became evident that the identity and reputation of the task requester has a significant influence on the task choice of some contributors. In addition to their own experiences, many contributors rely on collective assessments of requesters on the afore-mentioned Turkopticon platform.

[1] http://www.mturkgrid.com/ and http://turkernation.com/

D. Geiger, *Personalized Task Recommendation in Crowdsourcing Systems*,
Progress in IS, DOI 10.1007/978-3-319-22291-2_5

Browser extension for personalized HIT recommendations

As part of a research project on personalized task recommendations at the University of Mannheim, Germany, we have developed a browser extension for Mechanical Turk workers. The metacrowd extension helps you find interesting tasks by generating **recommendations based on your individual preferences and qualifications.**

Using metacrowd is simple: Install the extension, navigate to mturk.com, and log into metacrowd using the sidebar. Please take a minute to read the brief instructions. You will receive recommendations for interesting tasks while you browse Mechanical Turk as usual. The quality of your recommendations improves over time as you submit and rate more tasks and thus train the system.

The extension is completely free and does not require you to provide any personally identifiable information. Evaluations will be strictly anonymous and for academic purposes only. Please see the privacy policy for further details.

Chrome: https://chrome.google.com/webstore/detail/emegnidnhhdnjdomeobc...
Firefox: https://addons.mozilla.org/firefox/addon/metacrowd

We are looking forward to hearing about your experiences and to addressing any questions or concerns that you might have. We are also happy to take suggestions for future improvements, e.g., to accommodate different types of usage behavior.

Fig. 5.1 Exemplary forum announcement

In September 2014, Turkopticon contains about 170,000 individual reviews for roughly 750 requesters. The aggregated reputations of requesters are typically integrated into the standard task list on Mechanical Turk using a simple user script, which demonstrates the willingness of contributors to use third-party services in order to improve their experience with the platform. Since metacrowd presents recommended tasks in a separate interface, support for Turkopticon ratings was added to the browser extension as described in Sect. 4.6.1.

Several other improvements were made based on the received feedback, such as the wish to restrict the recommended tasks according to the number of available assignments and the reward per contribution. As opposed to the complex preference model that the recommender derives from a contributor's past interactions, these criteria represent generic restrictions of the set of candidate tasks and were thus implemented in the form of explicit filter options. The contributors also raised some issues with the presentation of recommended tasks and with the integration of metacrowd's UI elements into the Mechanical Turk website, which were subsequently addressed.

Finally, the pilot study highlighted the challenge of building a trusted relationship with the community of Mechanical Turk workers. Due to bad experiences with dishonest requesters that offer payments far below minimum wage or refuse to pay for valid contributions, some contributors have grown wary of interacting with parties they are not familiar with. Several concerns were raised about the use of the collected data for commercial purposes, such as selling them to requesters. In an attempt to mitigate these concerns, the privacy policy was refined (see Sect. 4.6.1) and the academic nature of the project was emphasized. These concerns also confirmed the requirement of allowing users to register with anonymous credentials.

5.2 Contributor Survey

Once all of the issues discovered in the pilot study were resolved, a large-scale exploratory survey was conducted among 500 Mechanical Turk workers with a twofold purpose. Firstly, the survey was designed to gather further insights into task search behavior across a variety of contributors and to compile an empirical account of the potential of better mechanisms. Secondly, the survey aimed at measuring the perceived utility of the metacrowd service from a pre-adoption perspective and, at the same time, served as a means to acquire users for an evaluation of the service in productive use. The following subsections discuss the survey development process, the data collection, and the subsequent data analysis.

5.2.1 Questionnaire Design

In line with the described aims, the survey instrument consisted of three parts. The first part contained general questions about the respondent's experience and activity on the Mechanical Turk platform. This comprised the number of months the respondent had been working on the platform and the average time spent per week. Respondents were asked to round up to the next full month and hour, respectively.

The second part inquired about the respondent's process of finding new tasks. It opened with a question about the frequency of finding tasks via different channels, such as the list of HITs or the keyword search on mturk.com, word of mouth in forums or chats, and any third-party scripts or extensions. Respondents were then asked to rate the importance of various criteria when looking for new tasks, in particular the type or topic of a task, the number of available assignments, the reward, and the requester. The available items for both questions were based on results from existing studies into task search on Mechanical Turk (Chilton et al. 2010; Schulze et al. 2011), on the considerations about contributor preferences in Sect. 1.1, and on further insights gained during the pilot study. Respondents were also given the possibility to rate additional items using an 'other' option. Finally, two questions inquired about the time spent to find new tasks and the relation to the time spent working on tasks.

The third part introduced the metacrowd service by providing links to the browser extension in the Chrome and Firefox stores and asked the respondents to assess its usefulness. The policies of the Mechanical Turk platform, which was used to recruit the survey respondents, prohibit requesters to ask contributors to install software or register with external services. Consequently, the respondents were only asked to examine the description and the screenshot of the browser extension as shown in Fig. 4.4 above. The instructions, however, also stated that the respondents were free to try out the extension and complete the survey within 12 hours.

The questionnaire design followed well-founded recommendations and guidelines aggregated in extensive reviews of the survey research methodology literature (Krosnick and Presser 2010; Krosnick et al. 2014), which are briefly summarized in the following. In order to minimize satisficing, i.e., the tendency to reduce the cognitive effort involved in providing high-quality answers, the survey was kept as short and motivating as possible. It included a welcome page with a short description of its purpose and started off with easy questions to encourage the respondents. Questions were grouped in the afore-mentioned parts in order to avoid the additional cognitive load of switching back and forth between topics. Several revisions ensured that the wording was easy to understand, unambiguous, and neutral to avoid leading or loaded questions.

An open-ended question format was used to measure the experience of contributors in the first part of the survey instead of offering a set of ranges, since any particular choice of options might have biased the respondents. When being asked to rate a list of individual items in the second part of the survey, the order of items was randomized to avoid primacy and recency effects. In general, a question format that allowed respondents to directly rate the underlying dimension of interest, such as importance, frequency, or usefulness, was preferred over 'agree-disagree' or 'true-false' statements as the latter tend to introduce bias by inducing an acquiescent response style. They also increase cognitive complexity by requiring the respondent to first answer a comparable rating question in their mind and then match this answer with the available options.

The corresponding rating scales implemented evidence-based recommendations on the number of response options, using five points for unipolar scales, which ask a respondent to rate one particular attribute, and seven points for bipolar scales, which ask a respondent to weigh two opposite attributes. All scale points were verbally labeled, which has been shown to maximize data quality and respondent satisfaction. Furthermore, 'no opinion' or 'don't know' response options were omitted as research shows that they do not lead to an improvement in data quality and may encourage satisficing. Finally, the rating scales included mandatory open-ended comment fields that required the survey participants to provide additional details on their responses. Besides enabling further insights into the collected data, the increased effort of providing meaningful textual responses also served as a means to discourage and detect deliberate cases of strong satisficing, i.e., respondents that invest little or no cognitive effort to complete the survey as quickly as possible.

The methodological literature also highlights the value of pretesting surveys, which was done in two phases. At first, the questionnaire was refined using a peer review approach. Multiple researchers provided input on the various questions and response scales. Once any identified issues were resolved, the questionnaire was subjected to a series of field pretests with members of the target population. Six rounds with 25 participants in total were conducted to optimize the layout and detect problems with individual survey items, such as subtle ambiguities in wording. For instance, the question "How much time do you spend looking for new HIT groups?" triggered several comments of the type "I rarely stop looking for new HITs" or "I am constantly looking for new hits when I'm not working on them." Consequently, the question was refined to "How much time do you spend to find a new HIT group?" in order to emphasize the focus on the isolated effort of finding the next task to work on. In another case, respondents kept adding the Turkopticon rating as an important criterion for task choice using the 'other' option, until it was clarified that the 'requester' item also refers to the corresponding reputation. The final questionnaire is depicted in Appendix D.

5.2.2 *Data Collection*

The survey was administered by means of a paid online questionnaire, which was hosted on the popular SurveyGizmo website. All participants from the target population of Mechanical Turk workers were recruited by publishing a corresponding task on the platform. The task paid a reward of 50–60 cents and, as customary, was open to all workers with a HIT approval rate greater than or equal to 90 %. The HIT approval rate denotes the proportion of accepted submissions and is used to exclude obviously fraudulent individuals. The description of the task encouraged both new and experienced contributors to participate and included a brief introduction with a link to the actual questionnaire. A screenshot of the task is displayed in Fig. 5.2.

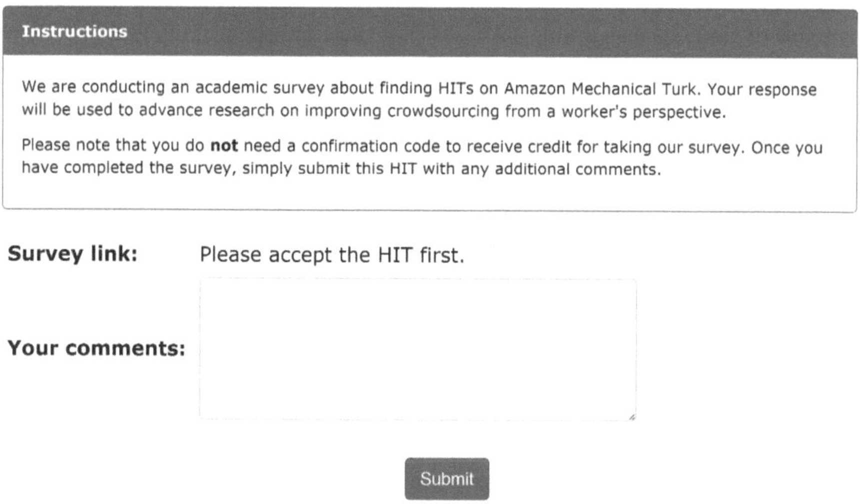

Fig. 5.2 Survey recruitment task

In order to sample a heterogeneous cross-section of contributors and to increase visibility in different parts of the world, the survey task was published in two series of batches à 30–50 assignments, which were spread over the course of a workday and of a weekend with several days in between. Within a series, a new batch was published every 3–4 hours and each batch remained on the platform until all assignments were completed or the survey was closed. A total of 490 complete responses were collected in 12 batches.

The collected responses were then cleansed of duplicates. Within each batch series, a filter prevented respondents from filling out the survey twice, using their unique Mechanical Turk worker id, which was submitted as a parameter in the survey link. Since the filter, however, did not take effect across the two series of batches, 14 individuals responded to the survey twice. In line with the intended filter behavior, only the first response was kept. Another four cases in which respondents violated Mechanical Turk policies by using two worker accounts in parallel or in immediate succession were identified using the IP address. As these were considered deliberate attempts at cheating, all eight responses were removed.

In a next step, a subset of the responses was flagged for inspection using several indicators for strong satisficing behavior. This included conspicuous comments, non-differentiation of ratings ("straight-lining"), and response times in the lower decile of one of the three survey parts (10, 97, and 48 seconds, respectively). Cases that matched at least one of the inspection criteria were manually checked for fake responses, which were characterized by implausible response times of only a few seconds, meaningless comments, or clear contradictions between ratings and comments. Out of 163 inspected cases, 26 low-quality responses were removed, which resulted in a remaining total of 442 responses.

Finally, all comments were scanned to identify individual problems with the corresponding questions. Seven respondents misinterpreted the term "HIT group" in question 5 and stated that they did not work on batch tasks. Regarding the same question, the comments of four other respondents indicated that they assessed the amount of time spent working on tasks, not the time spent finding them. Three respondents stated that they lacked the knowledge to answer the question. A few problems were also identified with respect to question 7. Two respondents stated that they could not access the metacrowd information due to technical problems. One respondent refused to "go to other sites for a survey." The corresponding partial responses to both questions were excluded from the analysis.

5.2.3 Data Analysis

In line with the targeted population, the experience and activity of the 442 respondents vary widely as illustrated in Fig. 5.3. A quarter of the respondents have spent less than 4 months on the Mechanical Turk platform, while another quarter have been long-term contributors for more than 2 years and up to 8 years. The median was located at 12 months. Similarly, the lower and upper quartiles for the average

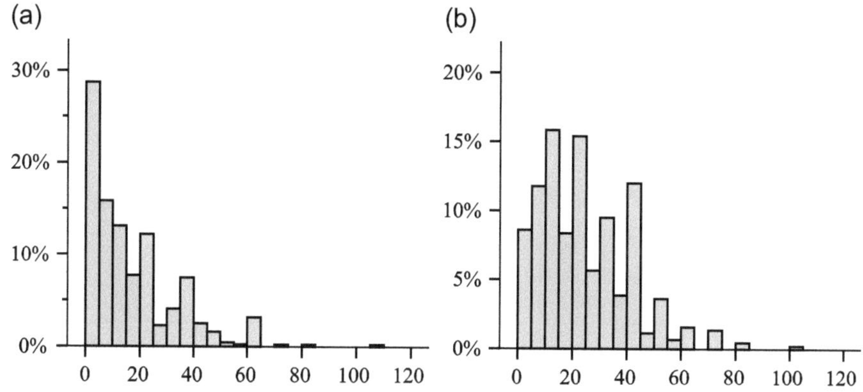

Fig. 5.3 Platform experience and activity frequency distributions. (**a**) Total months working on Mechanical Turk. (**b**) Weekly hours working on Mechanical Turk

weekly time working on Mechanical Turk were determined as 10 and 32.75 hours, respectively; the median was located at 20 hours per week. Consequently, it appears that about one fourth of the respondents use Mechanical Turk as a significant source of income.

The survey also collected paradata on the geographic location of the respondents and the used browser. The large majority of the respondents, 82.4 %, were located in the United States as determined by their IP address. A further 12.4 % of the responses came from India; the remainder was spread all over the globe. Based on the transmitted user agent strings, 62.9 % of the respondents used a Chrome browser to access the survey, while 27.4 % used Firefox. Less than 10 % used Microsoft Internet Explorer, Apple's Safari, or mobile browsers on iPads, iPhones, and Android devices. Note that some respondents might employ multiple browsers to work on Mechanical Turk.

5.2.3.1 Search Channels

The majority of the respondents make frequent use of the standard Mechanical Turk user interface to find tasks. As shown in Fig. 5.4, over 60 % of the respondents use the list of HITs and the built-in keyword search often or very often, while only about 15 % of the respondents stated that they rarely or never rely on these mechanisms. When it comes to finding tasks by word of mouth, the distribution is much more uniform: roughly 40 % of the respondents have never or rarely found tasks in that manner as opposed to 45 % who often or very often learn about interesting tasks from their fellow contributors. An analysis of the corresponding comments shows that these respondents regularly frequent external worker forums and various task lists on reddit.com. Finally, a quarter of the respondents rely often or very often, a further 10 % sometimes, on the use of third-party scripts or extensions. This includes mechanisms to extend Mechanical Turk's list of tasks with additional information, most importantly the Turkopticon ratings that were

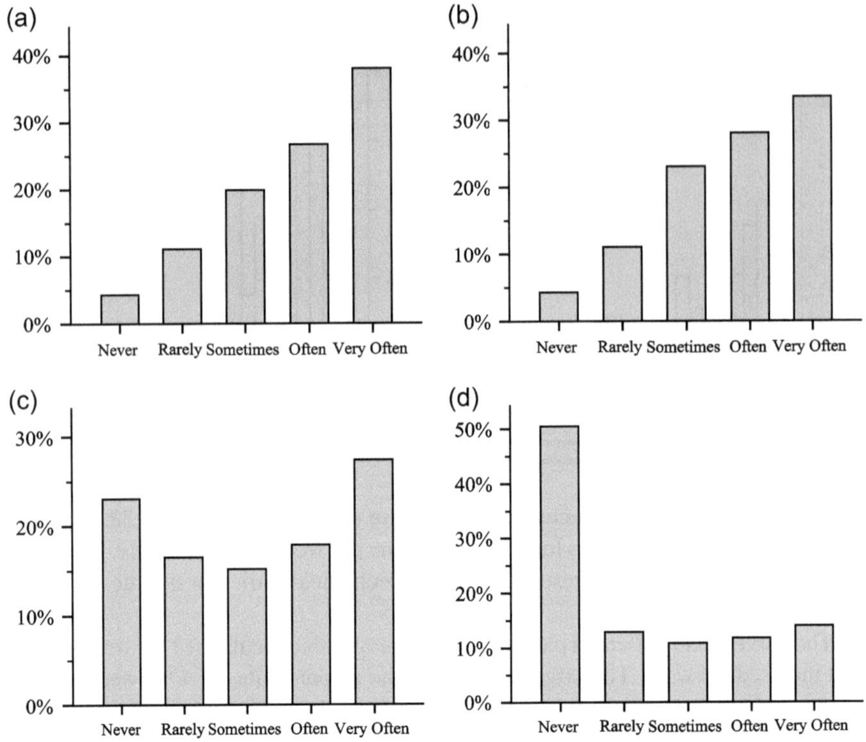

Fig. 5.4 Search channel frequency distributions. (**a**) Use of list of HITs. (**b**) Use of keyword search. (**c**) Use of word of mouth. (**d**) Use of third-party scripts or extensions

discussed in Sect. 5.1. Many respondents also use scripts and extensions that perform regular scans for updates on predefined keyword searches, such as 'HIT Scraper' or 'Page Monitor.'

A correlation analysis using the Spearman rank correlation coefficient (Spearman 1904) was conducted to uncover potential associations between the response variables. The correlation coefficient r_s measures the strength of the observed association as a value between 0 and 1. A corresponding p-value denotes the probability of observing this or a more extreme result if the null hypothesis (in this case: no monotonic correlation) were true. In contrast to the commonly used Pearson product-moment correlation, the Spearman correlation is defined on both continuous and ordinal scales. Moreover, it is robust towards outliers, and also measures non-linear, monotonic relationships. Table D.1 lists the calculated correlations for all survey items.

With respect to the different search channels, the results most notably show a strong positive correlation among the use of external mechanisms to find tasks, i.e., word of mouth and third-party scripts ($r_s = .468$, $p < .0005$). Individuals that rely on these two channels seem to make less use of standard platform interfaces, i.e., the list of tasks ($r_s = -.217$ and $r_s = -.161$, respectively) and the keyword search ($r_s = -.124$ and $r_s = -.180$, respectively), and vice versa. Moreover, such 'sophisticated' search behavior is more prominent among power users as indicated by positive associations between the use of external mechanisms and the average weekly hours on the platform ($r_s = .248$ and $r_s = .226$, $p < .0005$). No association was identified between the use of different task search mechanisms and the number of months a contributor has spent on the platform.

5.2.3.2 Search Criteria

Being asked about the criteria they look for in a task, over 40 % of the respondents state that the type or topic is very or extremely important to them. Close to 30 % declare it is moderately important, whereas about 30 % claim only a slight or no importance at all. The number of available assignments appears to bear a smaller degree of importance to many, although responses are similarly mixed: Almost half of the respondents consider the number of assignments as a criterion that is not or only slightly important, whereas close to 30 % attribute moderate and less than 25 % high or extreme importance. The variance in responses is considerably smaller when it comes to the reward offered for a task. Almost 90 % of the respondents see this as a very or extremely important criterion. Likewise, about 65 % state that the requesters and their reputation are very or extremely important to their task choice, with another 20 % claiming moderate importance. The corresponding frequency distributions are depicted in Fig. 5.5.

The correlation analysis in Table D.1 shows moderately strong associations between the importance of the requester and both the use of word of mouth channels ($r_s = .408$, $p < .0005$) and third-party scripts ($r_s = .242$, $p < .0005$). To a weaker degree, this effect can also be observed for the number of available assignments. These results correspond to the kind of recommendations available through word of mouth channels, which typically point to tasks from reputable requesters with a large amount of assignments. Similarly, the purpose of many of the commonly used third-party scripts is to highlight tasks from requesters with good reputations. Respondents that put an increased importance on the type or topic of a task, on the other hand, tend to rely less on word of mouth channels ($r_s = -.140$, $p = .003$) and more frequently on individual search processes using standard platform interfaces.

It also appears that respondents who spend more weekly time on the platform put an increased importance on the requester ($r_s = .257$, $p < .0005$), the number of available assignments ($r_s = .228$, $p < .0005$), and the size of the reward ($r_s = .136$, $p = .004$). Further research is required to determine the exact nature of the observed relationships between platform activity, task selection criteria, and preferred search channels.

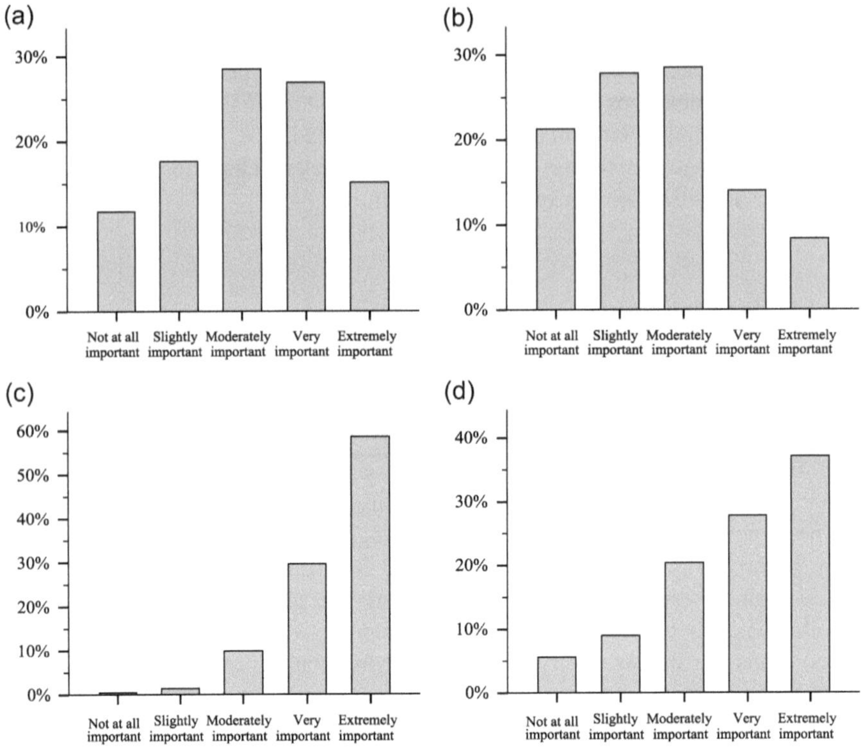

Fig. 5.5 Search criteria frequency distributions. (**a**) Importance of type or topic. (**b**) Importance of number of assignments. (**c**) Importance of reward. (**d**) Importance of requester

5.2.3.3 Search Time

Instead of asking respondents for an estimate of the average time they spend to find a task, the survey inquired about the frequency of five different time intervals on an ordinal scale. This style of questioning gave consideration to the possibility that individual search times, besides reflecting a basic tendency, might vary with alternating search strategies and situational factors. Estimating a single overall average would thus be cognitively complex and result in a loss of information. The assessment of these interval frequencies was complemented with an additional question that asked respondents, on a general level, to compare the time they spend finding new tasks to the time they spend working on tasks.

The use of multiple response items also enabled an analysis of the degree of non-differentiation as measured, for example, by the case-wise standard deviations σ of the five interval ratings. The analysis reveals that 7.5 % of the respondents did not differentiate between the intervals at all ($\sigma = 0$). Another 14 % used only two neighbored frequency options, mostly 'sometimes' and 'often,' for all five intervals ($0.4 \leq \sigma < 0.49$). An examination of the corresponding comments confirms the influence of situational factors. Many respondents highlighted that their search time varies with the overall range of tasks on a given day or time, the activity of their preferred requesters, or depending on whether they are busy working on something else at the same time.

Despite these external influences, the results show a wide variance in search time across individuals. The frequency distributions of the responses to the first question are depicted in a stacked bar chart in Fig. 5.6a. With respect to very fast search times, respondents can be divided into three approximately equal-sized groups that often or very often, sometimes, and never or rarely spend less than 20 seconds to find a new task. Search times between 20 seconds and 2 minutes are most common: About 45 % of the respondents often or very often spend that time to find tasks; another 40 % do this sometimes and about 15 % never or rarely. About 30 % of the respondents often or very often spend a time interval of 2–5 minutes to find a new task, whereas another 30 % sometimes and close to 40 % never or rarely spend this amount of time. Finally, about 25 % of the respondents often or very often, 20 % sometimes, and 55 % never or rarely spend more than 5 minutes.

One particularly interesting insight from these results, given the typically brief nature of assignments on the Mechanical Turk platform, is that a considerable amount of the respondents are regularly going to great lengths to find new tasks. This is also reflected in the responses to the second question on search time, whose distribution is depicted in Fig. 5.6b. While, as expected, the majority of respondents spend more or much more time working on the platform, a remarkably large fraction of 29 % claim that they require about the same or even more time to find tasks. Another 16.6 % state they only spend slightly more time on working on tasks. These results corroborate the findings of existing studies (Chilton et al. 2010) and strongly emphasize the opportunities for mechanisms that improve the search process.

The correlation analysis suggests that respondents who spend more weekly time on the platform more often invest long time intervals to find task, such as more than 5 minutes ($r_s = .245$, $p < .0005$). Longer search times were also positively

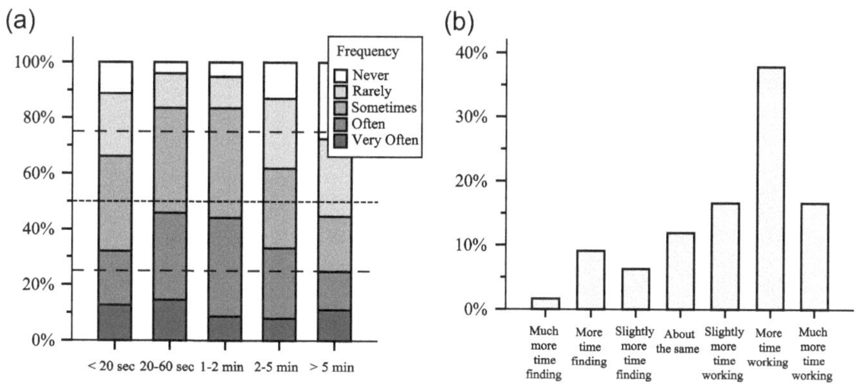

Fig. 5.6 Search time frequency distributions. (**a**) Time spent to find a new task. (**b**) Time spent finding new tasks compared to time spent working on tasks

related to the use of external search channels and to the degree of importance that a contributor attributes to the number of available task assignments. Future research will need to explore potential explanations for these relationships, such as an increased difficulty of finding suitable tasks for very active users or varying levels of efficiency associated with different search strategies.

5.2.3.4 Metacrowd

The reaction of the survey participants to the metacrowd service was predominantly positive. As the frequency distribution in Fig. 5.7 shows, almost half of the respondents believed that the extension would be very or even extremely useful to them. About 35 % found it moderately useful whereas about 20 % could see no or only a slight benefit for their personal task search.

An analysis of the comments revealed further details. Respondents that attributed a high usefulness to the service were enthusiastic about reducing their search effort and, at the same time, getting suggestions for tasks that match their personal interests and at which "they are good." Many participants liked the idea of filtering tasks by types and requesters. Several comments pointed out the benefits of preferring tasks from familiar requesters and filtering out unfair requesters. They also highlighted the integration of Turkopticon as a particularly useful feature. In line with these comments, the correlation analysis indicates a generally more positive rating by contributors who put an increased importance on the type or topic of a task ($r_s = .142$, $p = .003$) and on its requester ($r_s = .178$, $p < .0005$).

A number of respondents appreciated that new recommendations were pushed directly into a separate section of the Mechanical Turk environment, thus not requiring them to go to other pages or otherwise interfering with their regular

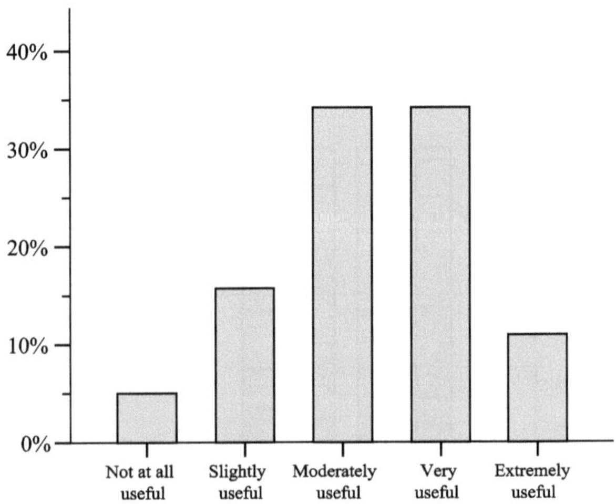

Fig. 5.7 Perceived usefulness frequency distribution

process. Several comments inquired about the frequency of updates given the quick disappearance rate of well-paid tasks. Some participants expressed concerns about security and privacy issues. Others worried that the extension might slow down their browser, would not consider qualifications, or not be able to adequately capture their preferences. Corresponding to the routine trade-off between search costs and preferences discussed in Sect. 1.1, a few respondents also suspected that they would need to improve their preference profile by explicitly rating tasks because their submission history would contain some tasks they did "just to keep busy and not because [they] like them."

5.2.4 *Discussion*

The results of this survey extend the knowledge base on the search behavior of contributors to the Amazon Mechanical Turk platform. The findings largely correspond to both earlier research (Chilton et al. 2010; Schulze et al. 2011) and insights gained during the pilot study. In line with the positive first reception of the browser extension, they confirm the potential for task recommendation mechanisms in general and the decisions taken with respect to the design of the metacrowd service.

The survey also identified several associations among different aspects of contributor search behavior that may be the object of further research. Future studies might want to differentiate between casual and power users, which seem to employ a significantly different search process. The responses also confirmed that more emphasis should be put on the requester as an important task selection criterion. Two particularly interesting avenues should be the study of potential causal relationships between different task selection criteria and the use of specific search channels as well as a more detailed analysis of the factors that influence individual search time.

Despite the large number of participants, a self-selection bias affects the external validity of the results. Given the recruitment procedure, it seems likely that a disproportionately high fraction of the respondents are contributors who prefer to work on surveys and paid tasks. Moreover, the visibility of the survey task may have varied with individual search preferences, such as the criteria used to sort the list of tasks. As with any survey that is based on a non-probability sample, the results may thus not be entirely representative with respect to the demographics, interests, and further characteristics of the Mechanical Turk worker population.

Finally, as explained in Sect. 5.2.1, the survey instructions asked the respondents to assess the potential utility of the metacrowd service by examining a description and a screenshot of the browser extension, but did not require them to install it. While the comments on the last question indicate that a number of participants had actually started using metacrowd before answering the question, the responses mostly capture an initial assessment of the service. Indeed, the collected paradata shows that 90 % of the respondents completed this last part of the survey within less than four minutes. In order to complement these results, the next section will evaluate the use of the service during regular contribution activity by analyzing data that was collected over the course of several months.

5.3 Online Evaluation

Following the pilot study, a revised prototype of the metacrowd system was deployed for a period of 109 days. During this time, the system recorded all relevant interactions between registered users and tasks by means of the tracking and synchronization mechanisms described in Sect. 4.6. For each examined or rejected recommendation, the system also stored additional information for evaluation purposes, such as the preference estimate and the number of data points used to calculate this estimate. The following analysis of this dataset first examines the overall activity of the monitored contributors and their use of the metacrowd service. In a second step, a closer examination of the recorded interactions with the recommender system is conducted to evaluate the system's predictive capability. The study concludes with a discussion of the challenges identified during the evaluation period and ways to address them in future versions.

5.3.1 Intensity of Use

During the evaluation period, 174 users downloaded the browser extension and registered for the metacrowd service. As suggested by the findings of the preceding survey, the data shows a broad spectrum of activity with respect to the Mechanical Turk platform. Figure 5.8 illustrates this variation in contribution behavior by depicting the size of the initially synchronized submission history per user (see Sect. 4.6.2) on a square root transformed scale. With the three quartiles at 26, 64, and 179 tasks, the distribution shows the presence of a number of power users but also highlights a large percentage of occasional or new users. The upper and lower deciles are located at 310 and 1.5 tasks, respectively.

A similarly heterogeneous picture is shown when analyzing the use of the metacrowd service. Individual intensity of use was measured as the number of recommendations with which a user interacted. Of the 174 registered users, a total of 118 examined or rejected at least one recommended task. As illustrated in Fig. 5.9, the evaluated dataset exhibits an equally wide variance across individuals. On the one hand, a quarter of the 118 users interacted with more than 33 recommended tasks each. The ten most active percent examined or rejected between 128 and 426 recommendations per user. On the other hand, the records also show interactions with less than 9 recommendations for half of the users.

No recommendation-related interactions were recorded for another 56 users. Judging from the size of their initial submission history and the number of ratings, it appears that about 30 % of them did not provide the system with enough data points

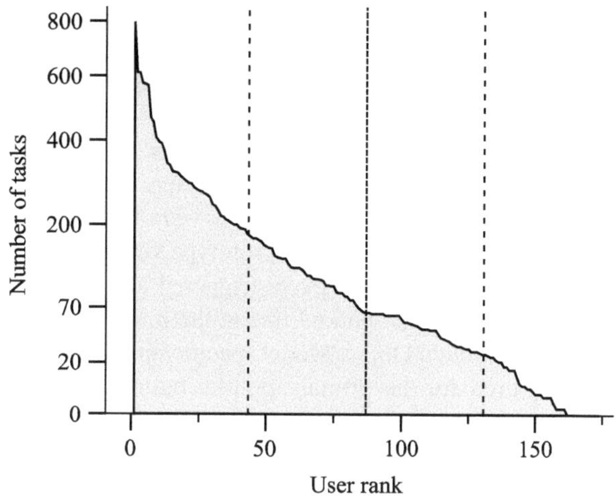

Fig. 5.8 Initial submission history size per user

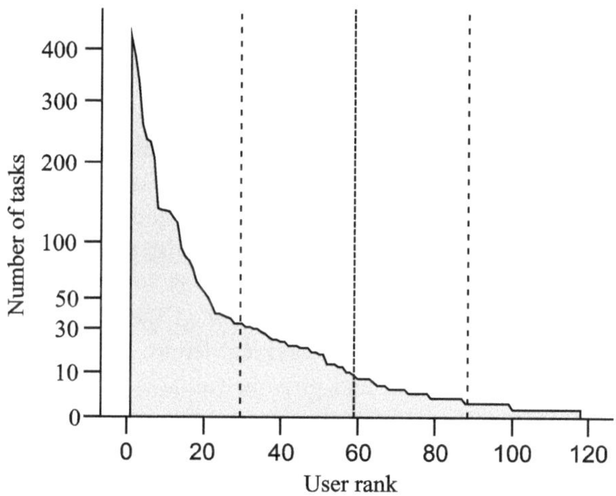

Fig. 5.9 Number of examined or rejected recommendations per user

to generate a single recommendation. In order to build the necessary knowledge of their preferences, these users would have needed to invest more effort in training the recommender system. It is also possible that some users prevented the synchronization of their submission history by logging out of Mechanical Turk before installing the extension or by using a different browser.

Moreover, two technical issues may have negatively influenced the evaluation. Firstly, the operation of the recommender service was interrupted several times due to outages of the Mturk-Tracker or overloads of the corresponding API, entailing further optimizations to the task synchronization adapter. These interruptions prevented the system from retrieving task updates for periods between a day and a week. During the first two outages, this might have led to the recommendation of outdated tasks that were not available for submission anymore. In order to address this issue, a mechanism was put in place that regularly closed all open tasks that had not received updates from Mturk-Tracker for more than one hour. During subsequent service outages in the evaluation period, metacrowd thus automatically stopped generating recommendations due to the lack of candidate tasks.

Secondly, a closer analysis of discrepancies in the proportions of the recorded interaction types revealed a malfunction of the tracking mechanism in certain environments. For about 15 % of the users in particular, a considerable number of ratings or rejected recommendations were tracked as opposed to only a few or no examined ones. The evaluated prototype version detected an examined recommendation by intercepting an event triggered at the user's click on the corresponding task link. While this worked fine at the time of testing, a recent refinement of the W3C's Document Object Model specification highlighted that a "click event should only be fired for the primary pointer button" (Kacmarcik et al. 2014). In newer browser versions, this appears to have resulted in breaking the tracking of recommendations that were examined in a new tab or window using, for example, the middle mouse button. Many individuals might prefer this way of interacting with the extension while working on a task or browsing the list of tasks in the current browser tab. As a consequence, the evaluated dataset may significantly underestimate the intensity of use.

Any examined recommendations that were not tracked would also not have disappeared from the sidebar list unless they were interacted with in another manner or not available anymore. Hence, the issue would lead to a rather static list of recommendations. A lack of responsiveness, just like multiple interruptions of the task synchronization, may have discouraged a number of users from engaging with the service for an extended period. It is also possible that further technical problems with individual environments, such as incompatibilities with specific browser versions or installed scripts, prevented the extension from working properly in some cases. Given the prototypical nature of the metacrowd service, such issues were to be expected and provide valuable insights for future development. The discussion in Sect. 5.3.3 suggests several avenues to address these challenges.

5.3.2 *Predictive Capability*

Recommender systems are often evaluated according to their capability to predict the true preferences of their users (Shani and Gunawardana 2011). Given a recommended item, these preferences manifest themselves most evidently in the user's reaction to this item. Corresponding to the two ways that a metacrowd user can interact with a task presented in the sidebar, the evaluated dataset contains a total of 2725 examined recommendations, reflecting positive user preferences, and 1802 rejected recommendations, reflecting negative preferences. Of the examined recommendations, 764 were eventually converted, i.e., the user submitted a contribution to the respective task. While explicit task ratings could provide an additional, more fine-granular differentiation of user preferences, these are only available for a subset of the examined recommendations. Moreover, their values strongly depend on individual interpretations of the rating scale and different rating behaviors (Desrosiers and Karypis 2011).

Assuming the recommender is able to accurately predict user preferences, a systematic effect should be observable between the strength of a prediction and the resulting user reaction. An intuitive indicator for the prediction strength is the calculated preference estimate, which is the prototype's primary means to select among the available candidate tasks. However, since the large majority of the available data on contributor preferences is derived from past submissions, most preference estimates vary closely around the predefined preference value that is assigned to all submission interactions (see Sect. 4.3.2). Moreover, since metacrowd only recommends tasks with a preference estimate above the mean of the rating scale, the dataset does not contain supposedly unfavorable recommendations. As a consequence, the per-user variance of the preference estimates in the evaluated dataset is too small to determine any statistically significant effects. While future experimental designs might deliberately increase the variance by injecting recommendations with a low preference estimate, such experiments are risky (Shani and Gunawardana 2011) and would most probably have resulted in a loss of user acceptance during the prototype evaluation.

Given the positive and homogeneous nature of the evaluated preference estimates, another well-suited indicator for the strength of a prediction is the number of data points on which the respective calculation is based. This measure equals the number of non-zero similarities between a candidate task and the contributor's set of (positively and negatively) preferred tasks and therefore varies substantially across both tasks and contributors. Consequently, this analysis focuses on testing the hypothesis that the number of data points, which can be perceived as the system's confidence in a prediction, is positively associated with a favorable user reaction.

A small number of ambiguous cases were excluded from the analysis. For example, three tasks were both examined and rejected, which might occur if a user was interacting with outdated lists of recommendations, such as in previously loaded browser tabs, or if two lists of recommendations were generated in quick succession. A further 17 tasks were rejected upon recommendation but submitted between a few minutes and several days later. Users may have rejected these tasks by accident or as a workaround if the system did not immediately remove them from the list of recommendations when being examined. The same reasoning applies to five more tasks, which received a very high rating shortly before or after rejection.

The method of analysis needs to take into account that the observations in the evaluated dataset are not entirely independent of each other. Whereas the recorded interactions are spread across a wide variety of tasks—each task is involved in a mean of 1.35 interactions and less than one percent of the tasks are a part of more than 5 interactions—they are heavily clustered on the respective users. Neglecting these user dependencies might bias the analysis and cause users with many observations to have a disproportionately high influence on the result. Consequently, the number of data points associated with the recommendations in each category of user reactions—examined, rejected, and converted ones—was averaged on a per-user basis. This step results in a set of three mean values for each user in the evaluated dataset. Moreover, the original data points scale was logarithmized to account for the fact that the marginal impact of an additional data point on the recommendation calculation decreases with larger numbers.

Figure 5.10a shows the mean per-user differences in the logarithmic numbers of data points for examined and rejected recommendations. In order to minimize contortion, only users for which both mean values are based on at least three recorded interactions were considered. A paired-samples *t*-test was conducted to determine any statistically significant mean difference across users between the two categories of interactions. Data are mean ± standard deviation, unless otherwise stated. There were no outliers in the data, as assessed by inspection of a boxplot. The difference scores for the examined and rejected recommendation means were normally distributed, as assessed by Shapiro-Wilk's test ($p = .359$) and visual inspection of a Normal Q-Q Plot. Across users, the mean logarithmic number of data points was higher for examined recommendations (1.252 ± 0.341) than for rejected ones (1.085 ± 0.354). This represents a statistically significant difference of 0.167 (95 % confidence interval, 0.079–0.254), $t(42) = 3.822$, $p < .0005$, where t denotes the test statistic. A common measure to quantify the size of the effect is given by Cohen's d, which is calculated as the standardized mean difference and yields a value of $d = .5829$ (Cohen 1988).

The observed effect is even more distinct when narrowing the comparison to the differences between rejected and converted recommendations, as shown in Fig. 5.10b. Again, the analysis only includes users for which the per-category means are each based on the values of at least three interactions. No outliers were present and the assumption of normality was not violated according to Shapiro-Wilk's test ($p = .071$) and visual inspection of a Normal Q-Q Plot. The mean logarithmic number of data points was higher for converted recommendations (1.392 ± 0.406) than for rejected ones (1.130 ± 0.293). This represents a

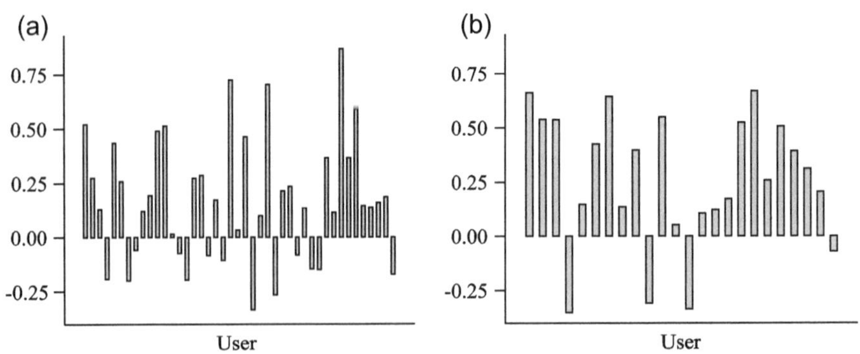

Fig. 5.10 Mean differences in the log number of data points per user. (**a**) Between examined and rejected recommendations. (**b**) Between converted and rejected recommendations

statistically significant difference of 0.262 (95 % confidence interval, 0.132–0.392), $t(23) = 4.155$, $p < .0005$, $d = .8482$.

In summary, this analysis of the interactions recorded during the evaluation period confirms the hypothesis that the recommender's confidence in a particular recommendation is positively associated with a favorable user reaction. Being highly statistically significant, the results strongly suggest that the observed examined and converted recommendations are not due to chance or a simple placebo effect but rather the result of metacrowd's ability to successfully predict its users' task preferences. Large effect sizes emphasize the practical relevance of the results.

5.3.3 Discussion

The prototype evaluation demonstrates that a third-party service such as metacrowd is a viable approach to implement and study task recommendation in the field. Despite its prototypical nature, a number of contributors have used the service extensively and benefited from the generated task recommendations. Overall, the results of this online experiment are highly encouraging as to the general acceptance and utility of personalized task recommendations. This is in line with the few existing studies in crowd creation systems (see Sect. 3.2.3). Future research should aim to validate these results in crowdsourcing systems from other domains and with different organizational functions.

On the other hand, the experiences gained during the evaluation have also highlighted the challenges that arise with the inherent external dependencies on the supported crowdsourcing platforms. While a screen scraping approach can be successfully applied to synchronize tasks and contributor data if the platform does not provide a stable API, it is naturally highly sensitive to changes in the user interface. In the case of Mechanical Turk, part of this burden was shifted by using the Mturk-Tracker as an intermediary, although this introduced an additional dependency. Unless a cooperation with the platform providers can be established, these challenges can only be approached by setting up an extensive testing and monitoring system in order to detect and adjust to changes as quickly as possible.

As stated above, the accuracy of the reported interaction frequencies is limited by the prototype's tracking and synchronization mechanisms. In order to improve reliability, these mechanisms require further refinement and ongoing validation in a variety of potential client environments. Given the discussed problem in intercepting user clicks in certain situations, the recorded numbers of examined and converted recommendations represent a lower bound on the actual values. The number of conversions might be further underestimated in cases in which the submission to an examined recommendation could not be tracked immediately but was later synchronized from the submission history. Due to the absence of task ids in Mechanical Turk's submission history, it is possible that the synchroni zation process could not retrieve a matching task from the Mturk-Tracker or that it marked a task similar to the examined one as submitted instead (see Sect. 4.6.2).

At the same time, the described issues may have led to an increased number of rejections if tasks appeared or remained on the list of recommendations although they had previously been examined or submitted. In other cases, users may have rejected recommendations because they were not eligible to work on the respective tasks. While the recommender considers proper qualification constraints during candidate task selection, some requesters use unstructured textual means instead to indicate that a task is, e.g., only available to the participants in a related task or to a group of explicitly invited workers.

Since the current implementation assumes that a rejected recommendation always reflects a mismatch with personal preferences, rejections for other reasons will bias the respective user profiles. In order to improve both data interpretability and recommendation behavior, future versions might want to collect additional information when removing recommendations from the list. This could be realized by introducing additional buttons or a follow-up question that inquires about the reasons for a rejection. In case a recommendation is actually rejected for preference reasons, this approach would also provide an opportunity for a more sophisticated content-based recommender to further inquire about the preferences for the individual features of the rejected task. Keeping the increased burden for the user in mind, the same idea could be applied to extend the task rating options.

Given the observed impact of confidence values on the recommender's predictive capability, it might be beneficial to reevaluate how to balance preference estimates and the number of data points when selecting and ordering candidate tasks. Furthermore, the communication of confidence values or even more detailed justifications for individual suggestions may provide more transparency on the recommendation process and thus increase the acceptance of the system (Herlocker et al. 2000). A higher level of transparency might also encourage users to rate more tasks and facilitate the implementation of exploration mechanisms that actively probe preferences in areas with low confidence (see Sect. 3.3.2). Such mechanisms could be an effective way to quickly learn about the preferences of new users but also to maintain the necessary diversity with long-term users.

Finally, by analyzing the capability of the recommender to accurately predict its users' preferences and reactions, this evaluation has focused on the utility of task recommendations to the contributors in a crowdsourcing system. Following the argument laid out in Sect. 1.2, attracting contributors with the right interests and capabilities also entails benefits for the task requesters, i.e., improving the outcome and the efficiency of the crowdsourcing approach. Future research might devise ways to specifically measure these benefits, which, in many cases, will require explicit assessments of contribution quality (see Sect. 3.3.3).

Chapter 6
Conclusion

This book has engaged in the study of personalized task recommendation mechanisms, which assist contributors in crowdsourcing systems with the discovery of tasks that match their individual interests and capabilities. In order to advance the body of knowledge regarding the potential and the design of such mechanisms, the previous chapters have described a series of consecutive studies, which are part of an overarching knowledge building process within the design science research paradigm. This concluding chapter provides a brief summary of the presented studies and highlights the major contributions.

Chapter 2 has introduced a conceptual framework for research into the crowdsourcing phenomenon from a socio-technical systems perspective. It establishes a well-founded definition of crowdsourcing systems as a special type of information systems and differentiates the functions that such systems can assume within an organizational environment, identifying four system archetypes with distinct characteristics. By channeling and relating otherwise isolated research, this framework enables a deeper and comparative understanding of different types of crowdsourcing systems and thus guides the systematic analysis and design of their components.

Applying the framework to the study of task recommendation approaches in crowdsourcing systems, Chap. 3 has presented a systematic review of the corresponding academic literature. The review connects the sparse research streams in various system types and has aggregated a comprehensive picture of the emerging state of personalized task recommendation. In addition to providing a foundation for the design of new task recommenders, the review has identified several issues for a future research agenda. This includes the need for more significant empirical results through large-scale online experiments, an improved dialog with mainstream recommender systems research, and the integration of various sources of contributor knowledge that exceed the boundaries of individual systems.

Building on this foundation, Chap. 4 has laid out a prototypical design for a third-party task recommendation service, which integrates with existing crowdsourcing systems. Metacrowd enhances these systems with task

D. Geiger, *Personalized Task Recommendation in Crowdsourcing Systems*,
Progress in IS, DOI 10.1007/978-3-319-22291-2_6

recommendation capabilities and provides a versatile platform to conduct corresponding field studies. It is built on a client-server architecture, which consists of a scalable backend and a browser extension that embeds directly into the user interface of the supported environments. This architecture enables metacrowd's recommendation engine to obtain the necessary data on contributors, tasks, and interactions through a set of client- and server-side synchronization mechanisms. While the prototype focuses on the well-established Mechanical Turk marketplace, metacrowd features a modular design that facilitates the integration with most other crowdsourcing systems.

Finally, Chap. 5 has provided empirical evidence on the potential of personalized task recommendation in real-world environments. A large-scale contributor survey on Mechanical Turk has confirmed the shortcomings of regular search processes, which materialize in long search times and substantial difficulties to find interesting tasks. The survey responses have also provided new insights into contributor behavior that warrant further research. This includes an association between the importance of specific search criteria and the use of particular search channels as well as the significantly different search behavior of very active contributors. At the same time, the survey acted as an instrument to recruit users for the deployed metacrowd prototype, presumably representing the first online evaluation of a task recommender in a crowd processing context. The observed usage of the prototype and a statistical analysis of its ability to predict contributor preferences have confirmed the general acceptance and the utility of the generated suggestions. In doing so, the evaluation has provided tangible evidence of metacrowd's usefulness as a platform for both the academic study and the practical implementation of personalized task recommendation in crowdsourcing systems.

Appendix A: Findings of Systematic Literature Review

Table A.1 Context and outlets of task recommendation studies

	Context	Outlet
Crowd processing		
Ambati et al. (2011)	Micro-task marketplaces	Workshop on Human Computation at the 25th AAAI Conf. on Artificial Intelligence
Yuen et al. (2011)	Micro-task marketplaces	Intl. Conf. on Internet of Things and 4th Intl. Conf. on Cyber, Physical and Social Computing
Yuen et al. (2012)	Micro-task marketplaces	19th Intl. Conf. on Neural Information Processing
Satzger et al. (2011)	Micro-task marketplaces	9th Intl. Conf. on Business Process Management
Ho and Vaughan (2012)	Micro-task assignment	26th AAAI Conf. on Artificial Intelligence
Difallah et al. (2013)	Micro-task assignment	22nd Intl. World Wide Web Conf.
Crowd solving		
Dror et al. (2011)	Question answering communities	17th ACM SIGKDD Intl. Conf. on Knowledge Discovery and Data Mining
Zhou et al. (2012)	Question answering communities	21st Intl. World Wide Web Conf.
Hu et al. (2008)	Question answering communities	2nd Intl. Conf. on Ubiquitous Information Management and Communication
Li and King (2010)	Question answering communities	19th ACM Intl. Conf. on Information and Knowledge Management
Kao et al. (2010)	Question answering communities	2010 ACM Symposium on Applied Computing
Riahi et al. (2012)	Question answering communities	21st Intl. Conf. on World Wide Web
Liu et al. (2010)	Question answering communities	11th Intl. Conf. on Web-Age Information Management

(continued)

D. Geiger, *Personalized Task Recommendation in Crowdsourcing Systems*,
Progress in IS, DOI 10.1007/978-3-319-22291-2

Table A.1 (continued)

	Context	Outlet
Liu et al. (2005)	Question answering communities	14th ACM Intl. Conf. on Information and Knowledge Management
Guo et al. (2008)	Question answering communities	17th ACM Conf. on Information and Knowledge Management
Yan and Zhou (2012)	Question answering communities	34th European Conf. on Information Retrieval
Li et al. (2011)	Question answering communities	20th ACM Intl. Conf. on Information and Knowledge Management
Qu et al. (2009)	Question answering communities	18th Intl. Conf. on World Wide Web
Liu and Hao (2012)	Question answering communities	Intl. Journal of Information and Education Technology
Zhu et al. (2011)	Question answering communities	5th Intl. Conf. on Knowledge Science, Engineering and Management
Castro-Herrera (2010)	Forums	3rd Intl. Workshop on Managing Requirements Knowledge
Zhou et al. (2009)	Forums	25th IEEE Intl. Conf. on Data Engineering
Horowitz and Kamvar (2010)	Social search	19th Intl. Conf. on World Wide Web
Crowd creation		
Cosley et al. (2007)	Wikipedia	12th Intl. Conf. on Intelligent User Interfaces
Yuan et al. (2009)	Wikipedia	Journal of Computer-Mediated Communication
Jamjoom et al. (2009)	Knowledge management	IBM Journal of Research and Development
Cosley et al. (2006)	MovieLens	SIGCHI Conf. on Human Factors in Computing Systems

Table A.2 Sources of task recommendation knowledge

	Individual contributor			Content		Collaborative
	Behavior	Opinions	Contributions	Task features	Contributions	Task interactions
Crowd processing						
Ambati et al. (2011)	Completed tasks			Description, timestamp, reward		
Yuen et al. (2011)	Completed tasks		Validity	Category, reward, allotted time		
Ho and Vaughan (2012)	Completed tasks		Validity	Category		
Satzger et al. (2011)	Completed tasks	Self-assessed skills	Validity	Tags (required skills)		
Difallah et al. (2013)	Completed tasks	Facebook profile		Description, category		
Yuen et al. (2012)	Completed, selected, and viewed tasks		Validity	Category		Interaction patterns
Crowd solving						
Dror et al. (2011)	Answered questions	Starred, tracked, 'answer-voted' questions; self-assessed interests	Rating	Description, category	Text of all answers	Interaction patterns
Zhou et al. (2012)	Answered questions	Starred questions	Rating	Description, category		
Horowitz and Kamvar (2010)	Answered questions	Self-/peer-assessed expertise, pref. requesters, external profiles	Rating	Description, requester		
Hu et al. (2008)	Answered questions	Refused recommendations	Rating	Description		
Li and King (2010)	Answered questions		Rating, content	Description		
Kao et al. (2010)	Answered questions		Rating, content	Description, category		

(continued)

Table A.2 (continued)

	Individual contributor			Content		Collaborative
	Behavior	Opinions	Contributions	Task features	Contributions	Task interactions
Riahi et al. (2012)	Answered questions		Rating	Description, tags		
Liu et al. (2010)	Answered questions		Rating	Description		
Liu et al. (2005)	Answered questions		Content	Description		
Castro-Herrera (2010)	Answered questions		Content	Description		Answering patterns
Zhou et al. (2009)	Answered questions		Content	Description	Text of all answers	
Guo et al. (2008)			Content	Description, category		
Yan and Zhou (2012)	Answered questions			Description, requester	Text of best answers	
Li et al. (2011)	Answered questions			Description, category		
Qu et al. (2009)	Answered questions			Description		
Liu and Hao (2012)	Answered questions			Description		
Zhu et al. (2011)	Answered questions			Tags		
Crowd creation						
Cosley et al. (2007)	Edited articles			Description (title), links		Editing patterns
Jamjoom et al. (2009)	Annotated processes					Annotation patterns
Cosley et al. (2006)		Rated movies				Rating patterns

Table A.3 Evaluation of task recommenders

	Experimental setting	Sample	Evaluated properties
Crowd processing			
Ambati et al. (2011)	Offline	NAACL 2010 workshop (5k contributors, 114 tasks)	Accuracy
Yuen et al. (2012)	Offline	NAACL 2010 workshop (1.6k contributors, 7.5k tasks, 3k categories)	Accuracy
Ho and Vaughan (2012)	Offline	Self-recruited (612 contributors, 8k tasks)	Accuracy
Yuen et al. (2011)	Offline	Self-recruited (12 contributors, 40 tasks, 14 each)	Accuracy
Difallah et al. (2013)	User study	Self-recruited (320 tasks)	Utility
Crowd solving			
Dror et al. (2011)	Offline	Yahoo! Answers (169k contributors, 1.3m question-answerer pairs)	Accuracy
Li and King (2010)	Offline	Yahoo! Answers (19k contributors, 19k questions, 54k answers)	Accuracy
Li et al. (2011)	Offline	Yahoo! Answers (243k contributors, 383k questions, 1.3m answers)	Accuracy
Qu et al. (2009)	Offline	Yahoo! Answers (61k contributors, 27k questions, 217k answers)	Accuracy
Zhou et al. (2012)	Offline	Yahoo! Answers (3.5k contributors, 1.3m answers)	Accuracy
Liu and Hao (2012)	Offline	Yahoo! Answers (217k questions)	Accuracy
Guo et al. (2008)	Offline	Yahoo! Answers (150k users, 119k questions, 773k answers)	Accuracy
Kao et al. (2010)	Offline	Yahoo! Answers (136 contributors, 200 questions)	Accuracy
Hu et al. (2008)	Offline	Yahoo! Answers (8 contributors, 700 questions)	Accuracy, coverage
Yan and Zhou (2012)	Offline	Yahoo! Answers (1.4k contributors, 1.9k questions, 5.4k question-answerer pairs), Tencent Wenwen (6.7k contributors, 20k questions, 61k question-answerer pairs)	Accuracy
Liu et al. (2005)	Offline	Wondir (852k question-answerer pairs)	Accuracy
Riahi et al. (2012)	Offline	Stackoverflow (22k contributors, 369k questions)	Accuracy
Liu et al. (2010)	Offline	Iask (138k contributors, 963k questions)	Accuracy
Zhu et al. (2011)	Offline	Tianya Wenda (265k users, 1.3m questions, 5.5m answers)	Accuracy
Castro-Herrera (2010)	Offline	Open source project forums (7 forums, 12k users, 25k questions, 95k posts)	Accuracy
Zhou et al. (2009)	Offline	Tripadvisor forum (40k users, 122k threads, 972k posts)	Accuracy, scalability

(continued)

Table A.3 (continued)

	Experimental setting	Sample	Evaluated properties
Crowd creation			
Cosley et al. (2007)	Online	Wikipedia (91 contributors, 3.1k recommendations)	Utility
Yuan et al. (2009)	Online	Wikipedia (6.6 k contributors, 2.2k adopters)	Utility
Cosley et al. (2006)	Online	Movielens (2.7k contributors, 437 adopters, 2.8k edits)	Utility

Appendix B: Recommender Performance Optimizations

In order to reflect the most recent knowledge about preferences and qualifications, metacrowd aims to generate dynamic, up-to-date recommendations that are recalculated every time a user interacts with the Mechanical Turk website. As illustrated in Sect. 4.3.2, recommendation generation is a data intensive process, which is based on a large number of task entities. Every invocation of the recommender service requires the system to load two sets of tasks: (i) the set of currently open tasks and (ii) the constantly growing contributor-specific set of tasks for which the system can derive a preference value.

Given an average number of about 2000 open tasks (see Sect. 1.1), a recommendation request for an exemplary long-time contributor with recorded preferences on a further 1000 tasks involves loading data on about 3000 tasks. Repeated measurements during development have shown that, without any application code optimizations, sample queries that retrieve this number of tasks from App Engine's distributed Datastore take up to 15 seconds to complete. In addition to the time factor, Datastore reads incur a monetary cost of $0.06 per 100,000 retrieved entities at the time of writing (Google 2014g). If a reasonably active user triggered about 100 recommendation calculations per day, once at every interaction with the platform, the average monthly Datastore costs would thus amount to more than $5 per user, not counting any other system operations.

Metacrowd addresses these challenges by implementing a layered caching architecture. At the infrastructure level, the system makes use of App Engine's distributed Memcache (see Sect. 4.5.3). Since the use of Memcache limits the need to access the Datastore to those tasks that are not present in the cache, it reduces the respective costs to a minor factor. If the cache is fully populated with all requested tasks, it also reduces the time required for the above-mentioned sample queries by several seconds, which is, however, still not fast enough for a live recommendation generation.

In order to further speed up the recommendation process, metacrowd implements additional in-memory caches in application code. These caches store static task properties that are relevant to recommendation generation, i.e., task keywords,

D. Geiger, *Personalized Task Recommendation in Crowdsourcing Systems*,
Progress in IS, DOI 10.1007/978-3-319-22291-2

requesters, and qualification constraints. Except for increased memory needs, they incur no additional costs and have negligible access times of a few milliseconds. As opposed to the distributed Memcache, in-memory caches are not shared across the instances of an App Engine application, which run in separate virtual machines and memory spaces. Recommendation generation thus usually takes more time on recently started instances when many of the relevant task properties have not yet been cached. Metacrowd reduces these waiting times by estimating the relevant tasks during instance start-up and pre-populating the caches with the corresponding values. While it cannot predict the contributors whose recommendation requests will be routed to a particular instance—which would be required to cache the tasks associated with past preferences—metacrowd will warm up the caches for the set of currently open tasks and thus cover all potential recommendation candidates.

In line with the large data requirements, recommendation generation is also a computationally intensive process. The number of similarity calculations required to determine all preference estimates for a particular contributor corresponds to the size of the Cartesian product of (i) the set of candidate tasks and (ii) the set of tasks for which a preference value is available. If all of the 2000 open tasks, on average, are considered as candidates, a recommendation request for a new user with only 50 recorded preferences requires the recommender to calculate 100,000 similarities. As the number of recorded preferences increases with every relevant interaction, recommendation requests for long-time contributors would involve calculating several million similarities.

Metacrowd minimizes the number of similarity calculations in two ways. Firstly, it only considers the 300 most recent relevant interactions per type (see Sect. 4.3.2), which creates an upper boundary of 900 task preferences. The actual number is typically smaller as a contributor may interact with the same task in multiple ways; for instance, submit a contribution followed by a rating. Secondly, metacrowd aims to narrow down the set of candidate tasks as early as possible to avoid the calculation of irrelevant similarities. In particular, it eagerly determines and removes the candidates for which a contributor does not have the proper qualifications, which is typically a large part of all open tasks. Moreover, any filters on the number of HITs or the size of the reward are also applied in the early stages of the recommendation process.

By applying all of the described optimizations, the average recommendation times during the prototype evaluation consistently stayed below one second. In case that an increased user load or more complex similarity calculations should warrant further optimizations, the caching of similarities would be a viable option. It is a general advantage of most item similarities that they are relatively static and can therefore be cached (Owen et al. 2012, p. 58). This is particularly true for the Mechanical Turk tasks whose keyword and requester properties do not change after creation. Reading a cached similarity value would, however, need to be considerably faster than recalculating it and could thus only be realized using instance caches. These caches, in turn, could be regularly synchronized across instances using the Memcache.

Appendix C: Unique Property Service

At the time of metacrowd's development, the non-relational Datastore available on the Google App Engine does not provide native support for unique non-key properties. Due to the Datastore's distributed nature, the entities that possess a particular property may be spread across several nodes. Although the Datastore maintains a global index of selected property values to serve corresponding queries, the delay between changes to a property value and the update of the global index prevents strongly consistent assertions on the uniqueness of a specific property value.

A solution to this problem was implemented along with the prototype. The *unique property service* leverages the Datastore's support for unique keys, for transactions that span several entity groups, and for strongly consistent reads by entity keys. For every unique value, the service maintains a separate *unique value entity* whose (unique) key is a combination of: (i) the kind of the entity that defines the unique property, hereafter called the *domain entity*, (ii) the name of the unique property, and (iii) a string representation of the unique value. This unique value entity also has a property that contains the Datastore key of the domain entity that holds the unique value. Figure C.1 illustrates this relationship by use of an example. From an abstract perspective, the collective of unique value entities forms an index of unique values, which can be accessed in a strongly consistent manner.

Whenever a domain entity is to be persisted, the unique property service performs a strongly consistent series of operations within the same transaction. For each unique property, the service checks whether a unique value entity with the described key pattern exists. If it does not, a new unique value entity will be associated with the key of the domain entity, and the transaction is committed. In case a corresponding unique value entity exists and is already associated with the current domain entity, the value of the unique property has not changed and no further action is required; the transaction is also committed. If a corresponding unique value entity exists but is associated with a different domain entity, the service verifies that the assignment is still valid by retrieving the associated domain entity and inspecting the value of the unique property. If the unique value is in fact taken, the entire transaction will be rolled back, thus maintaining the uniqueness

D. Geiger, *Personalized Task Recommendation in Crowdsourcing Systems*,
Progress in IS, DOI 10.1007/978-3-319-22291-2

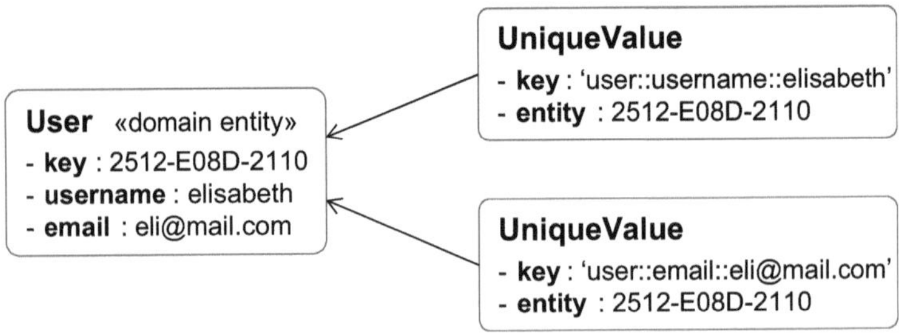

Fig. C.1 Exemplary unique value entities

constraint. In case the associated domain entity does not exist or does not have a property with the given name and value anymore, the unique value assignment is considered obsolete (see below). The existing unique value entity will then be reassociated with the key of the current domain entity and the transaction is committed.

The check for obsolete assignments is required since, at the time of validation, the unique property service does typically not know the previous value of the unique property, if any. Consequently, it cannot reconstruct the key of a previously assigned unique value entity and is unable to delete this entity. In order to avoid an overly pollution of the index with leftover assignments, future versions of the service might introduce a reverse way to look up the unique value entity that is assigned to a given domain entity and unique property, e.g., by use of an additional index.

In addition to maintaining uniqueness constraints, the unique property service can also be used to retrieve model entities by their unique property values in a strongly consistent manner—as opposed to a regular, eventually consistent query by the respective property. To this end, the service performs two consecutive reads: one to look up the respective unique value entity in the constructed index and another one to retrieve and return the associated domain entity if the unique value assignment is still valid. This effectively allows the use of unique properties as additional keys, similar to the notion of non-primary candidate keys in relational database schemas.

The unique property service includes a bean validation constraint annotation and a corresponding validator, which can easily be embedded in the regular validation process performed in the adapters. Note that, while the mechanism provides a very useful extension to the Datastore's capabilities, it is subject to several limitations: Firstly, the service must be invoked whenever a unique property is changed in order to properly update the index. Secondly, in order to avoid inconsistencies between the constructed index and the actual property values due to Datastore failures or race conditions, the service requires that both model entities and the corresponding unique value entities are updated within the same transaction. It is thus restricted by the Datastore limit of five entity groups per transactions, including one entity group per unique property. Thirdly, unique properties are limited to data types that can be represented as a unique string, since this string becomes part of the unique value entity's key. Finally, maintaining unique properties slightly increases the time required to store an entity, although this will be a negligible factor in most circumstances.

Appendix D: Contributor Survey

Thank you for participating in a short survey about finding HITs on Amazon Mechanical Turk. Your response will be used to advance research on improving crowdsourcing from a worker's perspective. These questions should take about 5 minutes of your time.

Fig. D.1 Contributor survey: welcome page

We'd first like to find out a little bit about your experience with Mechanical Turk.

1. How long have you been working on Mechanical Turk?
(Please round up to the nearest month.) *

month(s)

2. On average, how much time per week do you spend working on Mechanical Turk?
(Please round up to the nearest hour.) *

hour(s) per week

Fig. D.2 Contributor survey: part one (platform experience)

D. Geiger, *Personalized Task Recommendation in Crowdsourcing Systems*,
Progress in IS, DOI 10.1007/978-3-319-22291-2

Next, we'd like to ask you a few questions about finding new HIT groups to work on.

3. How do you look for new HIT groups?
(If necessary, please add further options.)

	Never	Rarely	Sometimes	Often	Very Often
List of HITs on mturk.com *	○	○	○	○	○
Search field on mturk.com *	○	○	○	○	○
Word of mouth (forums, chat, etc.) *	○	○	○	○	○
Third-party scripts or extensions *	○	○	○	○	○
Enter another option	○	○	○	○	○

Please provide some details on your search process. Include any scripts or extensions that you use. *

4. How important are the following criteria when you look for new HIT groups?
(If necessary, please add further options.)

	Not at all important	Slightly important	Moderately important	Very important	Extremely important
Type or topic *	○	○	○	○	○
Number of HITs available *	○	○	○	○	○
Reward *	○	○	○	○	○
Requester (includes reputation) *	○	○	○	○	○
Enter another option	○	○	○	○	○

Please provide some details on your response. *

5. How much time do you spend to find a new HIT group? *

	Never	Rarely	Sometimes	Often	Very Often
< 20 seconds *	○	○	○	○	○
20 - 60 seconds *	○	○	○	○	○
1 - 2 minutes *	○	○	○	○	○
2 - 5 minutes *	○	○	○	○	○
> 5 minutes *	○	○	○	○	○

Please provide some details on your response. *

6. Please compare the time you spend finding new HIT groups to the time you spend working on HIT groups. *

Much more time spent finding HITs	More time spent finding HITs	Slightly more time spent finding HITs	About the same	Slightly more time spent working on HITs	More time spent working on HITs	Much more time spent working on HITs
○	○	○	○	○	○	○

Fig. D.3 Contributor survey: part two (search behavior)

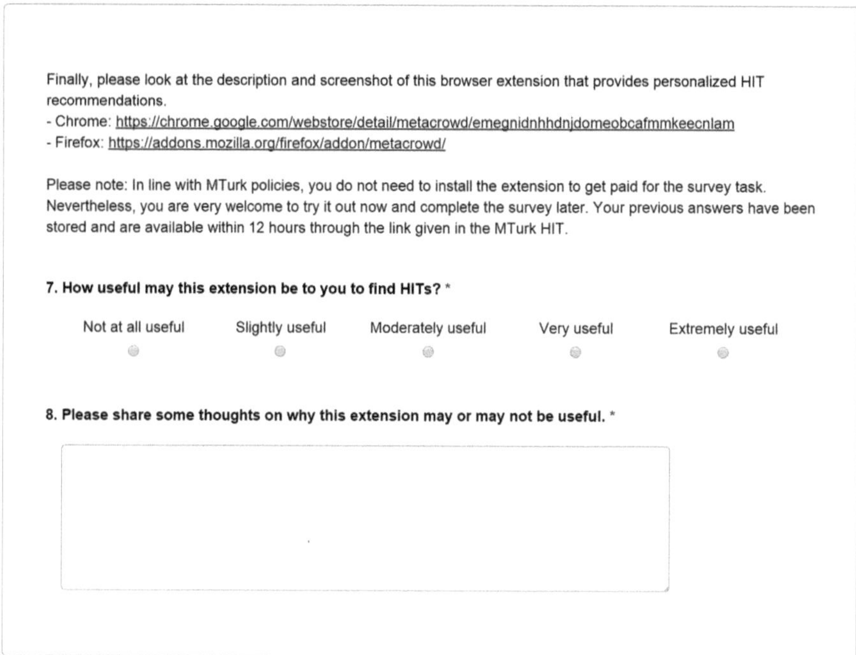
Finally, please look at the description and screenshot of this browser extension that provides personalized HIT recommendations.
- Chrome: https://chrome.google.com/webstore/detail/metacrowd/emegnidnhhdnjdomeobcafmmkeecnlam
- Firefox: https://addons.mozilla.org/firefox/addon/metacrowd/

Please note: In line with MTurk policies, you do not need to install the extension to get paid for the survey task. Nevertheless, you are very welcome to try it out now and complete the survey later. Your previous answers have been stored and are available within 12 hours through the link given in the MTurk HIT.

7. How useful may this extension be to you to find HITs? *

Not at all useful | Slightly useful | Moderately useful | Very useful | Extremely useful

8. Please share some thoughts on why this extension may or may not be useful. *

Fig. D.4 Contributor survey: part three (metacrowd)

Table D.1 Spearman correlations between contributor survey variables

		Platform activity		Search channels				Search criteria				Search time						Metacrowd
		TOT	WKLY	LIST	KWS	WOM	EXT	TYPE	REQ	NUM	REW	<20 s	20–60 s	1–2 min	2–5 min	>5 min	S/W	PU
Platform activity	TOT		.075	.023	.025	−.047	−.001	.050	−.053	−.002	.071	−.135**	−.020	.055	.057	.053	−.068	−.087
	WKLY	.075		.088	−.053	.248**	.226**	−.036	.257**	.228**	.136**	−.025	−.012	.138**	.186**	.245**	.114*	−.005
Search channels	LIST	.023	.088		.184**	−.217**	−.161**	.111*	.027	.074	.107*	.073	.074	.111*	.130**	.068	−.075	.123**
	KWS	.025	−.053	.184**		−.124**	−.180**	.121*	.039	.062	.100*	.007	.033	.004	.017	−.022	−.055	.116*
	WOM	−.047	.248**	−.217**	−.124**		.468**	−.140**	.408**	.175**	.033	.046	.059	.111*	.146**	.098*	.228**	.085
	EXT	−.001	.226**	−.161**	−.180**	.468**		−.065	.242**	.148**	.013	.073	.084	.018	.083	.113*	.137**	.009
Search criteria	TYPE	.050	−.036	.111*	.121*	−.140**	−.065		.086	.155**	.012	−.060	−.018	.120*	.052	.025	−.010	.142**
	REQ	−.053	.257**	.027	.039	.408**	.242**	.086		.200**	.120*	.114*	.031	.054	.071	.052	.076	.178**
	NUM	−.002	.228**	.074	.062	.175**	.148**	.155**	.200**		.059	.009	−.050	.074	.149**	.220**	.156**	.102*
	REW	.071	.136**	.107*	.100*	.033	.013	.012	.120*	.059		.103*	.029	.011	−.011	.008	−.048	.057
Search time	<20 s	−.135**	−.025	.073	.007	.046	.073	−.060	.114*	.009	.103*		.556**	−.073	−.271**	−.326**	−.198**	.073
	20–60 s	−.020	−.012	.074	.033	.059	.084	−.018	.031	−.050	.029	.556**		.195**	−.190**	−.297**	−.145**	.134**
	1–2 min	.055	.138**	.111*	.004	.111*	.018	.120*	.054	.074	.011	−.073	.195**		.527**	.310**	.135**	.066
	2–5 min	.057	.186**	.130**	.017	.146**	.083	.052	.071	.149**	−.011	−.271**	−.190**	.527**		.778**	.371**	.025
	>5 min	.053	.245**	.068	−.022	.098*	.113*	.025	.052	.220**	.008	−.326**	−.297**	.310**	.778**		.440**	−.043
	S/W	−.068	.114*	−.075	−.055	.228**	.137**	−.010	.076	.156**	−.048	−.198**	−.145**	.135**	.371**	.440**		−.089
Metacrowd	PU	−.087	−.005	.123**	.116*	.085	.009	.142**	.178**	.102*	.057	.073	.134**	.066	.025	−.043	−.089	

Key: *TOT* total months, *WKLY* weekly hours, *LIST* task list, *KWS* keyword search, *WOM* word of mouth, *EXT* scripts or extensions, *TYPE* type or topic, *REQ* requester, *NUM* number of assignments, *REW* reward, *S/W* ratio of time spent searching tasks to time spent working on tasks, *PU* perceived usefulness

** Correlation is significant at the 0.01 level (2-tailed)

* Correlation is significant at the 0.05 level (2-tailed)

$N \in [425; 442]$

References

Aciar, S., Zhang, D., Simoff, S., & Debenham, J. (2007). Informed recommender: Basing recommendations on consumer product reviews. *IEEE Intelligent Systems, 22*(3), 39–47. doi:10.1109/MIS.2007.55.

Ackoff, R. L. (1971). Towards a system of systems concepts. *Management Science, 17*(11), 661–671.

Ackoff, R. L. (1993). Presentation at the Systems Thinking in Action Conference, Cambridge, MA.

Allio, R. J. (2004). CEO interview: The InnoCentive model of open innovation. *Strategy & Leadership, 32*(4), 4–9. doi:10.1108/10878570410547643.

Alonso, O. (2011). *Perspectives on infrastructure for crowdsourcing* (pp. 7–10). In Workshop on Crowdsourcing for Search and Data Mining (CSDM 2011), Hong Kong, China.

Alter, S. (2008). Defining information systems as work systems: Implications for the IS field. *European Journal of Information Systems, 17*(5), 448–469. doi:10.1057/ejis.2008.37.

Ambati, V., Vogel, S., & Carbonell, J. (2011). *Towards task recommendation in micro-task markets* (pp. 80–83). Workshops at the 25th AAAI Conference on Artificial Intelligence, San Francisco, CA.

Antikainen, M., Mäkipää, M., & Ahonen, M. (2010). Motivating and supporting collaboration in open innovation. *European Journal of Innovation Management, 13*(1), 100–119. doi:10.1108/14601061011013258.

Balog, K., Azzopardi, L., & de Rijke, M. (2009). A language modeling framework for expert finding. *Information Processing & Management, 45*(1), 1–19. doi:10.1016/j.ipm.2008.06.003.

Becerra-Fernandez, I. (2000). The role of artificial intelligence technologies in the implementation of people-finder knowledge management systems. *Knowledge-Based Systems, 13*(5), 315–320.

Bennett, J., & Lanning, S. (2007). *The Netflix prize*. KDD Cup and Workshop, San Jose, CA.

Bernard, E. (2013). Bean validation 1.1. *JSR 349*. Available at http://jcp.org/en/jsr/detail?id=349

Bizer, C., Heath, T., & Berners-Lee, T. (2009). Linked data—The story so far. *International Journal on Semantic Web and Information Systems, 5*(3), 1–22. doi:10.4018/jswis.2009081901.

Brabham, D. C. (2008). Moving the crowd at iStockphoto: The composition of the crowd and motivations for participation in a crowdsourcing application. *First Monday, 13*(6), 1–22. doi:10.5210/fm.v13i6.2159.

Brabham, D. C. (2010). Moving the crowd at Threadless: Motivations for participation in a crowdsourcing application. *Information, Communication & Society, 13*(8), 1122–1145. doi:10.1080/13691181003624090.

Bray, T. (2014). The JavaScript Object Notation (JSON) data interchange format. *RFC 7159*. Internet Engineering Task Force. Available at http://www.rfc-editor.org/info/rfc7159

D. Geiger, *Personalized Task Recommendation in Crowdsourcing Systems*,
Progress in IS, DOI 10.1007/978-3-319-22291-2

Buitelaar, P., & Eigner, T. (2008). *Topic extraction from scientific literature for competency management* (pp. 55–66). 7th International Semantic Web Conference (ISWC), Karlsruhe, Germany.

Bulut, M. F., Yilmaz, Y. S., & Demirbas, M. (2011). *Crowdsourcing location-based queries* (pp. 513–518). IEEE International Conference on Pervasive Computing and Communications Workshops (PERCOM Workshops), Seattle, WA. doi:10.1109/PERCOMW.2011.5766944

Bunge, M. (2003). *Emergence and convergence: Qualitative novelty and the unity of knowledge.* Toronto: University of Toronto Press.

Burke, R. (2002). Hybrid recommender systems: Survey and experiments. *User Modeling and User-Adapted Interaction, 12*(4), 331–370. doi:10.1023/A:1021240730564.

Burke, R. (2007). Hybrid web recommender systems. In P. Brusilovsky, A. Kobsa, & W. Nejdl (Eds.), *The Adaptive Web: Lecture Notes in Computer Science* (pp. 377–408). Berlin: Springer. doi:10.1007/978-3-540-72079-9_12.

Burke, R., O'Mahony, M. P., & Hurley, N. J. (2011). Robust collaborative recommendation. In F. Ricci, L. Rokach, B. Shapira, & P. B. Kantor (Eds.), *Recommender systems handbook* (pp. 805–835). Boston: Springer. doi:10.1007/978-0-387-85820-3_25.

Burke, R., & Ramezani, M. (2011). Matching recommendation technologies and domains. In F. Ricci, L. Rokach, B. Shapira, & P. B. Kantor (Eds.), *Recommender systems handbook* (pp. 367–386). Boston: Springer. doi:10.1007/978-0-387-85820-3_11.

Buschmann, F., Meunier, R., Rohnert, H., Sommerlad, P., & Stal, M. (1996). *Pattern-oriented software architecture, volume 1: A system of patterns.* New York: Wiley.

Carignani, A., & Lorenzo, N. (2010). 'We don't need no education': Required abilities in online labor markets. 31st International Conference on Information Systems, St. Louis, MO.

Carpenter, J. (2011). May the best analyst win. *Science, 331*(6018), 698–699. doi:10.1126/science. 331.6018.698.

Carvalho, J. A. (2000). Information system? Which one do you mean? In E. D. Falkenberg, K. Lyytinen, & A. A. Verrijn-Stuart (Eds.), *Information system concepts: An integrated discipline emerging: IFIP—The International Federation for Information Processing* (pp. 259–277). Boston: Springer. doi:10.1007/978-0-387-35500-9_22.

Castro-Herrera, C. (2010). *A hybrid recommender system for finding relevant users in open source forums* (pp. 41–50). 3rd International Workshop on Managing Requirements Knowledge, Sydney. doi:10.1109/MARK.2010.5623811

Checkland, P. B. (1988). Information systems and systems thinking: Time to unite? *International Journal of Information Management, 8*(4), 239–248. doi:10.1016/0268-4012(88)90031-X.

Chilton, L. B., Horton, J. J., Miller, R. C., & Azenkot, S. (2010). Task search in a human computation market. In *Proceedings of the ACM SIGKDD Workshop on Human Computation (HCOMP)* (pp. 1–9). New York: ACM. doi:10.1145/1837885.1837889

Chua, S. J. (2007). Using Web 2.0 to locate expertise. In *Proceedings of the 2007 Conference of the Center for Advanced Studies on Collaborative Research (CASCON)* (pp. 284–287). New York: ACM. doi:10.1145/1321211.1321250

Chung, C.-Y., Hsu, P.-Y., & Huang, S.-H. (2013). ßP: A novel approach to filter out malicious rating profiles from recommender systems. *Decision Support Systems, 55*(1), 314–325. doi:10. 1016/j.dss.2013.01.020.

Churchman, C. W. (1968). *The systems approach.* New York: Delta.

Cockburn, A. (2000). *Writing effective use cases: Agile software development series.* Boston: Addison-Wesley.

Cockburn, A. (2005). *Ports and adapters architecture.* Retrieved December 1, 2014, from http:// alistair.cockburn.us/Hexagonal+architecture

Cohen, J. (1988). *Statistical power analysis for the behavioral sciences.* Hillsdale, NJ: Lawrence Erlbaum Associates.

Cooper, S., Khatib, F., Treuille, A., Barbero, J., Lee, J., Beenen, M., et al. (2010). Predicting protein structures with a multiplayer online game. *Nature, 466*(7307), 756–760. doi:10.1038/ nature09304.

Corney, J. R., Torres-Sanchez, C., Jagadeesan, A. P., & Regli, W. C. (2009). Outsourcing labour to the cloud. *International Journal of Innovation and Sustainable Development, 4*(4), 294–313. doi:10.1504/IJISD.2009.033083.

Cosley, D., Frankowski, D., Terveen, L., & Riedl, J. (2006). Using intelligent task routing and contribution review to help communities build artifacts of lasting value. In *Proceedings of the SIGCHI Conference on Human Factors in Computing Systems (CHI)* (pp. 1037–1046). New York: ACM. doi:10.1145/1124772.1124928

Cosley, D., Frankowski, D., Terveen, L., & Riedl, J. (2007). SuggestBot: Using intelligent task routing to help people find work in Wikipedia. In *Proceedings of the 12th International Conference on Intelligent User Interfaces* (pp. 32–41). New York: ACM. doi:10.1145/1216295.1216309

Cosley, D., Lam, S. K., Albert, I., Konstan, J. A., & Riedl, J. (2003). Is seeing believing? How recommender interfaces affect users' opinions. In *Proceedings of the SIGCHI Conference on Human Factors in Computing Systems (CHI)* (pp. 585–592). New York: ACM. doi:10.1145/642611.642713

Davis, G. B. (2000). Information systems conceptual foundations: Looking backward and forward. In R. Baskerville, J. Stage, & J. I. DeGross (Eds.), *Organizational and social perspectives on information technology: IFIP—The International Federation for Information Processing* (pp. 61–82). Boston: Springer. doi:10.1007/978-0-387-35505-4_5.

Demartini, G. (2007). *Finding experts using Wikipedia* (pp. 33–41). 2nd International ExpertFinder Workshop (FEWS), Busan, South Korea.

Deng, H., King, I., & Lyu, M. R. (2012). Enhanced models for expertise retrieval using community-aware strategies. *IEEE Transactions on Systems, Man, and Cybernetics, Part B: Cybernetics, 42*(1), 93–106. doi:10.1109/TSMCB.2011.2161980.

Desrosiers, C., & Karypis, G. (2011). A comprehensive survey of neighborhood-based recommendation methods. In F. Ricci, L. Rokach, B. Shapira, & P. B. Kantor (Eds.), *Recommender systems handbook* (pp. 107–144). Boston: Springer. doi:10.1007/978-0-387-85820-3_4.

Difallah, D. E., Demartini, G., & Cudré-Mauroux, P. (2013). *Pick-a-crowd: Tell me what you like, and i'll tell you what to do—A crowdsourcing platform for personalized human intelligence task assignment based on social networks* (pp. 367–377). Proceedings of the 22nd International World Wide Web Conference (WWW), Rio de Janeiro, Brazil.

Doan, A., Ramakrishnan, R., & Halevy, A. Y. (2011). Crowdsourcing systems on the World-Wide Web. *Communications of the ACM, 54*(4), 86–96. doi:10.1145/1924421.1924442.

Dondio, P., & Longo, L. (2011). Trust-based techniques for collective intelligence in social search systems. In N. Bessis & F. Xhafa (Eds.), *Next generation data technologies for collective computational intelligence: Studies in computational intelligence* (pp. 113–135). Berlin: Springer. doi:10.1007/978-3-642-20344-2_5.

Dror, G., Koren, Y., Maarek, Y., & Szpektor, I. (2011). I want to answer; who has a question? Yahoo! answers recommender system. In *Proceedings of the 17th ACM SIGKDD International Conference on Knowledge Discovery and Data Mining (KDD)* (pp. 1109–1117). New York: ACM. doi:10.1145/2020408.2020582

Ein-Dor, P., & Segev, E. (1993). A classification of information systems: Analysis and interpretation. *Information Systems Research, 4*(2), 166–204. doi:10.1287/isre.4.2.166.

Estellés-Arolas, E., & González-Ladrón-de-Guevara, F. (2012). Towards an integrated crowdsourcing definition. *Journal of Information Science, 38*(2), 189–200. doi:10.1177/0165551512437638.

Evans, E. (2004). *Domain-driven design: Tackling complexity in the heart of software*. Boston: Addison-Wesley.

Falkenberg, E. D., Hesse, W., Lindgreen, P., Nilsson, B. E., Oei, J. L. H., Rolland, C., et al. (1998). *A framework of information systems concepts* (The FRISCO Report). International Federation for Information Processing.

Ferstl, O. K., & Sinz, E. J. (2008). *Grundlagen der Wirtschaftsinformatik* (6th ed.). München: Oldenbourg Wissenschaftsverlag.

Fielding, R. T. (2000). *Architectural styles and the design of network-based software architectures*. Doctoral dissertation, University of California, Irvine.

Fowler, M. (2003). *AnemicDomainModel*. Retrieved December 1, 2014, from http://www.martinfowler.com/bliki/AnemicDomainModel.html

Fowler, M. (2004). *Inversion of control containers and the dependency injection pattern*. Retrieved December 1, 2014, http://martinfowler.com/articles/injection.html

Freed, N., & Borenstein, N. (1996). Multipurpose internet mail extensions (MIME) part two: Media types. *RFC 2046*, Internet Engineering Task Force. Available at http://www.rfc-editor.org/info/rfc2046

Freyne, J., Coyle, L., Smyth, B., & Cunningham, P. (2010). Relative status of journal and conference publications in computer science. *Communications of the ACM, 53*(11), 124. doi:10.1145/1839676.1839701.

Geiger, D., Rosemann, M., Fielt, E., & Schader, M. (2012). *Crowdsourcing information systems—Definition, typology, and design*. 33rd International Conference on Information Systems, Orlando, FL.

Geiger, D., Rosemann, M., & Fielt, E. (2011a). *Crowdsourcing information systems—A systems theory perspective*. 22nd Australasian Conference on Information Systems, Sydney, Australia.

Geiger, D., & Schader, M. (2014). Personalized task recommendation in crowdsourcing information systems—Current state of the art. *Decision Support Systems, 65*, 3–16. doi:10.1016/j.dss.2014.05.007.

Geiger, D., Seedorf, S., Schulze, T., Nickerson, R., & Schader, M. (2011b). *Managing the crowd: Towards a taxonomy of crowdsourcing processes*. 17th Americas Conference on Information Systems, Detroit, MI.

Glass, R. L., & Vessey, I. (1995). Contemporary application-domain taxonomies. *IEEE Software, 12*(4), 63–76. doi:10.1109/52.391837.

Google. (2013). *App engine: Effective memcache*. Retrieved December 1, 2014, from http://developers.google.com/appengine/articles/scaling/memcache

Google. (2014a). *App engine: Java runtime environment*. Retrieved December 1, 2014, from http://developers.google.com/appengine/docs/java/

Google. (2014b). *App engine: Java datastore API*. Retrieved December 1, 2014, from http://developers.google.com/appengine/docs/java/datastore/

Google. (2014c). *App engine: Structuring data for strong consistency*. Retrieved December 1, 2014, from http://developers.google.com/appengine/docs/java/datastore/structuring_for_strong_consistency

Google. (2014d). *App engine: Memcache Java API*. Retrieved December 1, 2014, from http://developers.google.com/appengine/docs/java/memcache/

Google. (2014e). *App engine: Task queue Java API*. Retrieved December 1, 2014, from http://developers.google.com/appengine/docs/java/taskqueue/

Google. (2014f). *Chrome: Content scripts*. Retrieved December 1, 2014, from http://developer.chrome.com/extensions/content_scripts

Google. (2014g). *App engine: Pricing*. Retrieved December 1, 2014, from http://cloud.google.com/appengine/pricing

Gopal, R., Marsden, J. R., & Vanthienen, J. (2011). Information mining—Reflections on recent advancements and the road ahead in data, text, and media mining. *Decision Support Systems, 51*(4), 727–731. doi:10.1016/j.dss.2011.01.008.

Gray, P. (2005). *Manager's guide to making decisions about information systems*. Hoboken, NJ: Wiley.

Guo, J., Xu, S., Bao, S., & Yu, Y. (2008). Tapping on the potential of Q&A community by recommending answer providers. In *Proceedings of the 17th ACM Conference on Information and Knowledge Management* (pp. 921–930). New York: ACM. doi:10.1145/1458082.1458204

Hacker, S., & von Ahn, L. (2009). Matchin: Eliciting user preferences with an online game. In *Proceedings of the 27th International Conference on Human Factors in Computing Systems (CHI)* (pp. 1207–1216). New York: ACM. doi:10.1145/1518701.1518882

Hayes, C., Avesani, P., & Veeramachaneni, S. (2007). An analysis of the use of tags in a blog recommender system. In *Proceedings of the 20th International Joint Conference on Artificial Intelligence* (pp. 2772–2777). Menlo Park, CA: AAAI Press.

Heinrich, L. J., Heinzl, A., & Riedl, R. (2011). *Wirtschaftsinformatik: Einführung und Grundlegung* (4th ed.). Berlin: Springer.

Herlocker, J. L., Konstan, J. A., Terveen, L. G., & Riedl, J. T. (2004). Evaluating collaborative filtering recommender systems. *ACM Transactions on Information Systems, 22*(1), 5–53. doi:10.1145/963770.963772.

Herlocker, J., Konstan, J., & Riedl, J. (2000). Explaining collaborative filtering recommendations. In *Proceedings of the 2000 ACM Conference on Computer Supported Cooperative Work (CSCW)* (pp. 241–250). New York: ACM. doi:10.1145/358916.358995

Hevner, A., & Chatterjee, S. (2010). *Design research in information systems: Integrated series in information systems*. Boston: Springer. doi:10.1007/978-1-4419-5653-8.

Hevner, A. R., March, S. T., Park, J., & Ram, S. (2004). Design science in information systems research. *MIS Quarterly, 28*(1), 75–105.

Ho, C.-J., & Vaughan, J. W. (2012). *Online task assignment in crowdsourcing markets*. 26th AAAI Conference on Artificial Intelligence, Toronto, Canada.

Hommes, C. H. (2006). Heterogeneous agent models in economics and finance. In *Handbook of computational economics* (vol. 2, pp. 1109–1186). Amsterdam: Elsevier. doi:10.1016/S1574-0021(05)02023-X

Horowitz, D., & Kamvar, S. D. (2010). The anatomy of a large-scale social search engine. In *Proceedings of the 19th International World Wide Web Conference (WWW)* (pp. 431–440). New York: ACM. doi:10.1145/1772690.1772735

Howe, J. (2006). The rise of crowdsourcing. *Wired Magazine, 14*(6), 1–4.

Howe, J. (2009). *Crowdsourcing: Why the power of the crowd is driving the future of business*. New York: Crown Business.

Hrastinski, S., Edenius, M., Kviselius, N. Z., & Ozan, H. (2012). How can software support open innovation? Extending community and marketplace perspectives. *International Journal of Networking and Virtual Organisations, 10*(1), 1–17. doi:10.1504/IJNVO.2012.045208.

Hu, D., Gu, S., Wang, S., Wenyin, L., & Chen, E. (2008). Question recommendation for user-interactive question answering systems. In *Proceedings of the 2nd International Conference on Ubiquitous Information Management and Communication* (pp. 39–44). New York: ACM. doi:10.1145/1352793.1352803

Huber, M. W., Piercy, C. A., & McKeown, P. G. (2006). *Information systems: Creating business value*. Hoboken, NJ: Wiley.

Ipeirotis, P. G. (2010). Analyzing the Amazon Mechanical Turk marketplace. *XRDS, 17*(2), 16–21. doi:10.1145/1869086.1869094.

Irani, L. C., & Silberman, M. S. (2013). Turkopticon: Interrupting worker invisibility in Amazon Mechanical Turk. In *Proceedings of the SIGCHI Conference on Human Factors in Computing Systems (CHI)* (pp. 611–620). New York: ACM. doi:10.1145/2470654.2470742

Jamjoom, H., Qu, H., Buco, M., Hernandez, M., Saha, D., & Naghshineh, M. (2009). Crowdsourcing and service delivery. *IBM Journal of Research and Development, 53*(6), 1–10. doi:10.1147/JRD.2009.5429040.

Jeppesen, L. B., & Lakhani, K. R. (2010). Marginality and problem-solving effectiveness in broadcast search. *Organization Science, 21*(5), 1016–1033. doi:10.1287/orsc.1090.0491.

Jesic, D., Kovacevic, J., & Stankovic, M. (2011). Web technologies for open innovation. In *Proceedings of the 3rd International Conference on Web Science (WebSci)*. New York: ACM. doi:10.1145/2527031.2527062

Kacmarcik, G., Leithead, T., Rossi, J., Schepers, D., Höhrmann, B., Le Hégaret, P., et al. (2014). *Document object model (DOM) level 3 events specification*. W3C Working Draft. Retrieved December 1, 2014, from http://www.w3.org/TR/DOM-Level-3-Events/

Kao, W.-C., Liu, D.-R., & Wang, S.-W. (2010). Expert finding in question-answering websites: A novel hybrid approach. In *Proceedings of the 2010 ACM Symposium on Applied Computing* (pp. 867–871). New York: ACM. doi:10.1145/1774088.1774266

Karau, S. J., & Williams, K. D. (1993). Social loafing: A meta-analytic review and theoretical integration. *Journal of Personality and Social Psychology, 65*(4), 681–706. doi:10.1037/0022-3514.65.4.681.

Kaufmann, N., Schulze, T., & Veit, D. (2011). *More than fun and money. Worker motivation in crowdsourcing—A study on Mechanical Turk.* 17th Americas Conference on Information Systems, Detroit, MI.

Kern, R., Thies, H., Bauer, C., & Satzger, G. (2010). Quality assurance for human-based electronic services: A decision matrix for choosing the right approach. In F. Daniel & F. M. Facca (Eds.), *Current trends in Web engineering: Lecture notes in computer science* (pp. 421–424). Berlin: Springer. doi:10.1007/978-3-642-16985-4_39.

Kern, R., Thies, H., Zirpins, C., & Satzger, G. (2012). Dynamic and goal-based quality management for human-based electronic services. *International Journal of Cooperative Information Systems, 21*(1), 3–29. doi:10.1142/S0218843012400011.

Kim, H.-N., El-Saddik, A., & Jo, G.-S. (2011). Collaborative error-reflected models for cold-start recommender systems. *Decision Support Systems, 51*(3), 519–531. doi:10.1016/j.dss.2011.02.015.

Kleemann, F., Voß, G., & Rieder, K. (2008). Un(der)paid innovators: The commercial utilization of consumer work through crowdsourcing. *Science, Technology & Innovation Studies, 4*(1), 5–26.

Kolari, P., Finin, T., Lyons, K., & Yesha, Y. (2008). *Expert search using internal corporate blogs.* Workshop on Future Challenges in Expertise Retrieval (SIGIR), Singapore.

Kroenke, D. (2011). *Using MIS* (5th ed.). Prentice Hall PTR: Upper Saddle River, NJ.

Krosnick, J. A., Lavrakas, P. J., & Kim, N. (2014). Survey research. In H. T. Reis & C. M. Judd (Eds.), *Handbook of research methods in social and personality psychology* (2nd ed., pp. 404–442). Cambridge: Cambridge University Press.

Krosnick, J. A., & Presser, S. (2010). Question and questionnaire design. In P. V. Marsden & J. D. Wright (Eds.), *Handbook of survey research* (2nd ed., pp. 263–314). Bingley: Emerald Group Publishing.

Kulkarni, A., Gutheim, P., Narula, P., Rolnitzky, D., Parikh, T., & Hartmann, B. (2012). MobileWorks: Designing for quality in a managed crowdsourcing architecture. *IEEE Internet Computing, 16*(5), 28–35. doi:10.1109/MIC.2012.72.

Land, F. (1985). Is an information theory enough? *The Computer Journal, 28*(3), 211–215. doi:10.1093/comjnl/28.3.211.

Leimeister, J. M., Huber, M., Bretschneider, U., & Krcmar, H. (2009). Leveraging crowdsourcing: Activation-supporting components for IT-based ideas competition. *Journal of Management Information Systems, 26*(1), 197–224. doi:10.2753/MIS0742-1222260108.

Levenshtein, V. I. (1966). Binary codes capable of correcting deletions, insertions and reversals. *Soviet Physics Doklady, 10*(8), 707–710.

Levy, Y., & Ellis, T. J. (2006). A systems approach to conduct an effective literature review in support of information systems research. *Informing Science Journal, 9*, 181–212.

Li, B., & King, I. (2010). Routing questions to appropriate answerers in community question answering services. In *Proceedings of the 19th ACM International Conference on Information and Knowledge Management* (pp. 1585–1588). New York: ACM. doi:10.1145/1871437.1871678

Li, B., King, I., & Lyu, M. R. (2011). Question routing in community question answering: Putting category in its place. In *Proceedings of the 20th ACM International Conference on Information and Knowledge Management* (pp. 2041–2044). New York: ACM. doi:10.1145/2063576.2063885

Ling, K., Beenen, G., Ludford, P., Wang, X., Chang, K., Li, X., et al. (2005). Using social psychology to motivate contributions to online communities. *Journal of Computer-Mediated Communication, 10*, 4. doi:10.1111/j.1083-6101.2005.tb00273.x.

Lintott, C. J., Schawinski, K., Slosar, A., Land, K., Bamford, S., Thomas, D., et al. (2008). Galaxy Zoo: Morphologies derived from visual inspection of galaxies from the Sloan Digital Sky Survey. *Monthly Notices of the Royal Astronomical Society, 389*(3), 1179–1189. doi:10.1111/j.1365-2966.2008.13689.x.

Liu, X., Croft, W. B., & Koll, M. (2005). Finding experts in community-based question-answering services. In *Proceedings of the 14th ACM International Conference on Information and Knowledge Management (CIKM)* (pp. 315–316). New York: ACM. doi:10.1145/1099554.1099644

Liu, G., & Hao, T. (2012). User-based question recommendation for question answering system. *International Journal of Information and Education Technology, 2*(3), 243–246.

Liu, M., Liu, Y., & Yang, Q. (2010). Predicting best answerers for new questions in community question answering. In L. Chen, C. Tang, J. Yang, & Y. Gao (Eds.), *Proceedings of the 11th International Conference on Web-Age Information Management (WAIM)* (pp. 127–138). Berlin: Springer. doi:10.1007/978-3-642-14246-8_15.

Lopez, M., Vukovic, M., & Laredo, J. (2010). PeopleCloud service for enterprise crowdsourcing. In *Proceedings of the 2010 I.E. International Conference on Services Computing (SCC)* (pp. 538–545). Miami, FL: IEEE. doi: 10.1109/SCC.2010.74

Lops, P., de Gemmis, M., & Semeraro, G. (2011). Content-based recommender systems: State of the art and trends. In F. Ricci, L. Rokach, B. Shapira, & P. B. Kantor (Eds.), *Recommender systems handbook* (pp. 73–105). Boston: Springer. doi:10.1007/978-0-387-85820-3_3.

Ludford, P. J., Cosley, D., Frankowski, D., & Terveen, L. (2004). Think different: Increasing online community participation using uniqueness and group dissimilarity. In *Proceedings of the 2004 conference on Human factors in computing systems—(CHI)* (pp. 631–638). New York: ACM. doi:10.1145/985692.985772

Lyytinen, K., & Newman, M. (2008). Explaining information systems change: A punctuated socio-technical change model. *European Journal of Information Systems, 17*(6), 589–613. doi:10.1057/ejis.2008.50.

Maheswaran, M., Braun, T. D., & Siegel, H. J. (1999). Heterogeneous distributed computing. In J. G. Webster (Ed.), *Wiley encyclopedia of electrical and electronics engineering* (pp. 679–690). New York: Wiley. doi:10.1002/047134608X.W5716.

Marinho, L. B., Hotho, A., Jäschke, R., Nanopoulos, A., Rendle, S., Schmidt-Thieme, L., et al. (2012). *Recommender systems for social tagging systems*. Boston: Springer. doi:10.1007/978-1-4614-1894-8.

Markines, B., Stoilova, L., & Menczer, F. (2006). Bookmark hierarchies and collaborative recommendation. In *Proceedings of the 21st National Conference on Artificial Intelligence (AAAI)* (vol. 2, pp. 1375–1380). Boston: AAAI Press.

Martin, R. C. (1996). The dependency inversion principle. *C++ Report, 8*(6), 61–66.

Martin, R. C. (2012). *The clean architecture*. Retrieved December 1, 2014, from http://blog.8thlight.com/uncle-bob/2012/08/13/the-clean-architecture.html

Mason, W., & Watts, D. J. (2009). Financial incentives and the 'performance of crowds'. *ACM SIGKDD Explorations Newsletter, 11*(2), 100–108. doi:10.1145/1809400.1809422.

Morita, M., & Shinoda, Y. (1994). Information filtering based on user behavior analysis and best match text retrieval. In W. B. Croft & C. J. van Rijsbergen (Eds.), *Proceedings of the 17th Annual International ACM SIGIR Conference on Research and Development in Information Retrieval* (pp. 272–281). London: Springer. doi: 10.1007/978-1-4471-2099-5_28

Mozilla. (2014). *Content scripts*. Retrieved December 1, 2014, from http://developer.mozilla.org/Add-ons/SDK/Guides/Content_Scripts

Ni, X., Lu, Y., Quan, X., Wenyin, L., & Hua, B. (2012). User interest modeling and its application for question recommendation in user-interactive question answering systems. *Information Processing & Management, 48*(2), 218–233. doi:10.1016/j.ipm.2011.09.002.

Nickerson, R. C., Varshney, U., & Muntermann, J. (2013). A method for taxonomy development and its application in information systems. *European Journal of Information Systems, 22*(3), 336–359. doi:10.1057/ejis.2012.26.

Niwa, S., & Honiden, S. (2006). Web page recommender system based on Folksonomy Mining for ITNG'06 submissions. In *Proceedings of the 3rd International Conference on Information Technology: New Generations (ITNG)* (pp. 388–393). Las Vegas, NV: IEEE. doi:10.1109/ITNG.2006.140

Osmani, A. (2012). *Learning JavaScript design patterns.* Sebastopol, CA: O'Reilly Media.

Osmani, A. (2013). *Developing Backbone.js applications.* Beijing: O'Reilly Media.

Owen, S., Anil, R., Dunning, T., & Friedman, E. (2012). *Mahout in action.* Shelter Island, NY: Manning.

Palermo, J. (2008). *The onion architecture: Part 1.* Retrieved December 1, 2014, from http://jeffreypalermo.com/blog/the-onion-architecture-part-1/

Paolacci, G., Chandler, J., & Ipeirotis, P. G. (2010). Running experiments on Amazon Mechanical Turk. *Judgment and Decision Making, 5*(5), 411–419.

Pazzani, M. J., & Billsus, D. (2007). Content-based recommendation systems. In P. Brusilovsky, A. Kobsa, & W. Nejdl (Eds.), *The adaptive Web* (pp. 325–341). Berlin: Springer. doi:10.1007/978-3-540-72079-9_10.

Peffers, K., Tuunanen, T., Rothenberger, M. A., & Chatterjee, S. (2007). A design science research methodology for information systems research. *Journal of Management Information Systems, 24*(3), 45–77. doi:10.2753/MIS0742-1222240302.

Pericas-Geertsen, S., & Potociar, M. (2013). JAX-RS 2.0: The Java API for RESTful Web services. *JSR 339.* Available at http://jcp.org/en/jsr/detail?id=339

Piramuthu, S., Kapoor, G., Zhou, W., & Mauw, S. (2012). Input online review data and related bias in recommender systems. *Decision Support Systems, 53*(3), 418–424. doi:10.1016/j.dss.2012.02.006.

Provos, N., & Mazières, D. (1999). *A future-adaptable password scheme.* USENIX Annual Technical Conference, Monterey, CA.

Qu, M., Qiu, G., He, X., Zhang, C., Wu, H., Bu, J., & Chen, C. (2009). Probabilistic question recommendation for question answering communities. In *Proceedings of the 18th International World Wide Web Conference (WWW)* (pp. 1229–1230). New York: ACM. doi:10.1145/1526709.1526942

Rainer, R. K., & Cegielski, C. G. (2010). *Introduction to information systems* (3rd ed.). Hoboken, NJ: Wiley.

Riahi, F., Zolaktaf, Z., Shafiei, M., & Milios, E. (2012). Finding expert users in community question answering. In *Proceedings of the 21st International World Wide Web Conference (WWW)* (pp. 791–798). New York: ACM. doi:10.1145/2187980.2188202

Ricci, F., Rokach, L., & Shapira, B. (2011). Introduction to recommender systems handbook. In F. Ricci, L. Rokach, B. Shapira, & P. B. Kantor (Eds.), *Recommender systems handbook* (pp. 1–35). Boston: Springer. doi:10.1007/978-0-387-85820-3_1.

Ross, J., Irani, L., Silberman, M. S., Zaldivar, A., & Tomlinson, B. (2010). Who are the crowdworkers. In *Extended Abstracts on Human Factors in Computing Systems (CHI)* (pp. 2863–2872) ACM. doi:10.1145/1753846.1753873

Rubens, N., Kaplan, D., & Sugiyama, M. (2011). Active learning in recommender systems. In F. Ricci, L. Rokach, B. Shapira, & P. B. Kantor (Eds.), *Recommender systems handbook* (pp. 735–767). Boston: Springer. doi:10.1007/978-0-387-85820-3_23.

Ryan, R., & Deci, E. (2000). Intrinsic and extrinsic motivations: Classic definitions and new directions. *Contemporary Educational Psychology, 25*(1), 54–67. doi:10.1006/ceps.1999.1020.

Sabherwal, R., & King, W. R. (1995). An empirical taxonomy of the decision-making processes concerning strategic applications of information systems. *Journal of Management Information Systems, 11*(4), 177–214.

Satzger, B., Psaier, H., Schall, D., & Dustdar, S. (2011). Stimulating skill evolution in market-based crowdsourcing. In S. Rinderle-Ma, F. Toumani, & K. Wolf (Eds.), *Business process management, Proceedings of the 9th International Conference on BPM* (Lecture notes in computer science, pp. 66–82). Berlin: Springer. doi:10.1007/978-3-642-23059-2_9.

Schenk, E., & Guittard, C. (2011). Towards a characterization of crowdsourcing practices. *Journal of Innovation Economics, 7*(1), 93–107. doi:10.3917/jie.007.0093.

Schroer, J., & Hertel, G. (2009). Voluntary engagement in an open web-based encyclopedia: Wikipedians and why they do it. *Media Psychology, 12*(1), 96–120. doi:10.1080/15213260802669466.

Schulze, T., Krug, S., & Schader, M. (2012). *Workers' task choice in crowdsourcing and human computation markets.* 33rd International Conference on Information Systems, Orlando, FL.

Schulze, T., Seedorf, S., Geiger, D., Kaufmann, N., & Schader, M. (2011). *Exploring task properties in crowdsourcing—An empirical study on Mechanical Turk.* 19th European Conference on Information Systems, Helsinki, Finland.

Shani, G., & Gunawardana, A. (2011). Evaluating recommendation systems. In F. Ricci, L. Rokach, B. Shapira, & P. B. Kantor (Eds.), *Recommender systems handbook* (pp. 257–297). Boston: Springer. doi:10.1007/978-0-387-85820-3_8.

Silver, M. S., Markus, M. L., & Mathis Beath, C. (1995). The information technology interaction model: A foundation for the MBA core course. *MIS Quarterly, 19*(3), 361–390.

Spearman, C. (1904). The proof and measurement of association between two things. *The American Journal of Psychology, 15*(1), 72–101. doi:10.2307/1412159.

Stankovic, M., Breitfuss, W., & Laublet, P. (2011a). Linked-data based suggestion of relevant topics. In *Proceedings of the 7th International Conference on Semantic Systems (I-Semantics)* (pp. 49–55). New York: ACM. doi:10.1145/2063518.2063525

Stankovic, M., Jovanovic, J., & Laublet, P. (2011b). Linked data metrics for flexible expert search on the open Web. In G. Antoniou, M. Grobelnik, E. Simperl, B. Parsia, D. Plexousakis, P. Leenheer, & J. Pan (Eds.), *The Semantic Web: Research and applications* (Lecture notes in computer science, pp. 108–123). Berlin: Springer. doi:10.1007/978-3-642-21034-1.

Su, X., & Khoshgoftaar, T. M. (2009). A survey of collaborative filtering techniques. *Advances in Artificial Intelligence.* doi: 10.1155/2009/421425

Sun, Y., Fang, Y., & Lim, K. H. (2012). Understanding sustained participation in transactional virtual communities. *Decision Support Systems, 53*(1), 12–22. doi:10.1016/j.dss.2011.10.006.

Surowiecki, J. (2005). *The wisdom of crowds.* New York: Anchor Books.

Valacich, J., Schneider, C., & Behl, R. (2011). *Information systems today: International version* (5th ed.). Upper Saddle River, NJ: Pearson Education.

Vernon, V. (2013). *Implementing domain-driven design.* Upper Saddle River, NJ: Addison-Wesley.

Von Ahn, L., Maurer, B., McMillen, C., Abraham, D., & Blum, M. (2008). reCAPTCHA: Human-based character recognition via web security measures. *Science, 321*(5895), 1465–1468. doi:10.1126/science.1160379.

von Bertalanffy, L. (1972). The history and status of general systems theory. *The Academy of Management Journal, 15*(4), 407–426. doi:10.2307/255139.

Vukovic, M. (2009). Crowdsourcing for enterprises. In *Proceedings of the 2009 World Conference on Services—I* (pp. 686–692). Los Angeles, CA: IEEE. doi:10.1109/SERVICES-I.2009.56

Wang, G. A., Jiao, J., Abrahams, A. S., Fan, W., & Zhang, Z. (2013). ExpertRank: A topic-aware expert finding algorithm for online knowledge communities. *Decision Support Systems, 54*(3), 1442–1451. doi:10.1016/j.dss.2012.12.020.

Weber, R. (1997). *Ontological foundations of information systems.* Melbourne: Coopers & Lybrand and the Accounting Association of Australia and New Zealand.

Webster, J., & Watson, R. (2002). Analyzing the past to prepare for the future: Writing a literature review. *MIS Quarterly, 26*(2), xiii–xxiii.

WKWI. (1994). Profil der Wirtschaftsinformatik (Mitteilungen der Wissenschaftlichen Kommission Wirtschaftsinformatik im VHB). *Wirtschaftsinformatik, 36*(1), 80–81.

Wu, H., Zubair, M., & Maly, K. (2006). Harvesting social knowledge from Folksonomies. In *Proceedings of the 17th Conference on Hypertext and Hypermedia* (pp. 111–114). New York: ACM. doi:10.1145/1149941.1149962

Yan, Z., & Zhou, J. (2012). A new approach to answerer recommendation in community question answering services. In R. Baeza-Yates, A. de Vries, H. Zaragoza, B. Cambazoglu, V. Murdock, R. Lempel, & F. Silvestri (Eds.), *Advances in information retrieval, Proceedings of the 34th European Conference on Information Retrieval* (Lecture notes in computer science, pp. 121–132). Berlin: Springer. doi:10.1007/978-3-642-28997-2_11.

Yuan, Y. C., Cosley, D., Welser, H. T., Xia, L., & Gay, G. (2009). The diffusion of a task recommendation system to facilitate contributions to an online community. *Journal of Computer-Mediated Communication, 15*(1), 32–59. doi:10.1111/j.1083-6101.2009.01491.x.

Yuen, M.-C., King, I., & Leung, K.-S. (2011). Task matching in crowdsourcing. In *Proceedings of the International Conference on Internet of Things and 4th International Conference on Cyber, Physical and Social Computing* (pp. 409–412). Dalian: IEEE. doi:10.1109/iThings/CPSCom.2011.128

Yuen, M.-C., King, I., & Leung, K.-S. (2012). TaskRec: Probabilistic matrix factorization in task recommendation in crowdsourcing systems. In T. Huang, Z. Zeng, C. Li, & C. S. Leung (Eds.), *Proceedings of the 19th International Conference on Neural Information Processing* (Lecture notes in computer science, pp. 516–525). Berlin: Springer. doi:10.1007/978-3-642-34481-7_63.

Zhang, H., Horvitz, E., Miller, R. C., & Parkes, D. C. (2011). *Crowdsourcing general computation.* Workshop on Crowdsourcing and Human Computation at the 29th Conference on Human Factors in Computing Systems (CHI), Vancouver, BC, Canada.

Zheng, H., Li, D., & Hou, W. (2011). Task design, motivation, and participation in crowdsourcing contests. *International Journal of Electronic Commerce, 15*(4), 57–88. doi:10.2753/JEC1086-4415150402.

Zhou, Y., Cong, G., Cui, B., Jensen, C. S., & Yao, J. (2009). Routing questions to the right users in online communities. In *Proceedings of the 25th IEEE International Conference on Data Engineering* (pp. 700–711). Shanghai: IEEE. doi:10.1109/ICDE.2009.44

Zhou, T. C., Lyu, M. R., & King, I. (2012). A classification-based approach to question routing in community question answering. In *Proceedings of the 21st International World Wide Web Conference (WWW)* (pp. 783–790). New York: ACM. doi:10.1145/2187980.2188201

Zhu, H., Chen, E., & Cao, H. (2011). Finding experts in tag based knowledge sharing communities. In H. Xiong & W. B. Lee (Eds.), *Proceedings of the 5th International Conference on Knowledge Science, Engineering and Management* (pp. 183–195). Berlin: Springer. doi:10.1007/978-3-642-25975-3_17.

Zwass, V. (2010). Co-creation: Toward a taxonomy and an integrated research perspective. *International Journal of Electronic Commerce, 15*(1), 11–48. doi:10.2753/JEC1086-4415150101.

MIX
Papier aus verantwortungsvollen Quellen
Paper from responsible sources
FSC® C105338
FSC
www.fsc.org

If you have any concerns about our products,
you can contact us on
ProductSafety@springernature.com

In case Publisher is established outside the EU,
the EU authorized representative is:
Springer Nature Customer Service Center GmbH
Europaplatz 3, 69115 Heidelberg, Germany

Printed by Libri Plureos GmbH
in Hamburg, Germany